Emission free energy from the deserts

With my compliments!

Paul van Son

Paul van Son
Thomas Isenburg

Emission free energy from the deserts

How a 'crazy Desertec idea' has become reality in North Africa and the Middle East

Smart Book Publishers, The Hague 2019

Paul van Son and Thomas Isenburg 2019

ISBN 978-94-92460-26-4
nur 961

Publisher: Theo Oskam
Design: Wouter Fris
Cover photo: Thomas Isenburg - *Plant Noor 2 in Ouarzazate, Morocco*

This is a publication of Smart Book Publishers
www.smartbookpublishers.com

Content

Part 1

0. Introduction

0.1 *The Desertec initiative, once a love baby of the German industry*

The idea that the deserts of the Middle East and North Africa (MENA) could deliver enough renewable energy to cover the demand of their own population and to power a part of the European market was not entirely new in 2009, but in that year quite unexpectedly a group of the largest German companies launched a unique and sensational initiative to make that happen by 2050. That was very surprising, as at that time not many industrial players worldwide would give much credit to large plans for the development of 'outrageously expensive' renewables at all, left alone in politically sensitive areas such as North Africa and the Middle East. Despite much scepticism and exactly in the depressed era of a major financial crisis the so-called 'Desertec Industrial Initiative' (Dii) was founded in Munich, near the headquarters of giant multinationals Siemens and MunichRe, to work with the international community. It was a powerful sign of hope and spirit. This new venture became a clear mandate to bring the public and private sectors in the related countries in motion for the benefit of 'power from the deserts'. The expectation of the initiators in those days was that up to 15 percent of the European electricity demand could by 2050 be served with power from mainly 'huge solar thermal installations' in the deserts and electricity being transported via high voltage DC cables from the southern to the northern regions of the Mediterranean and from there to Germany. The Desertec vision quickly became very popular in Germany and far beyond, although it was not yet very clear at that time what it was really all about. The industry group Dii was meant to challenge the vision and to pave the way for implementation in a realistic and pragmatic way. The main motivation of the German industry originated, evidently, from the creation of a new market for profitable power plant projects in the deserts.

The idea in those days was, that such projects could only be financed if the electricity would be delivered physically to Europe, even to Germany, which was used to massive de facto subsidization of renewables. However, in due course Dii has placed the movement for power from the deserts in a broader international, intercultural, technological and energy market context. The initial focus became more orientated on localisation of the industry with benefits for desert countries and integration of not only 'green electricity', but in a broader sense 'green energy' from MENA into the global energy markets. To date hydrogen and other synthetic emission-free energy carriers are becoming part of the 'package'.

Desertec may initially have raised attention as a sort of 'top-down planned economy type of road map', imposed by mainly German companies. However, the industry group Dii has taken steps to make this false perception more realistic, more international. Today, the main players in Dii are the Chinese 'master of power grids', State Grid Corporation of China, the Saudi-based market leader in the region, ACWA Power, and German renewable energy leader, Innogy. Also, Siemens, ABB, First Solar and many international companies are (again) on board. In the region itself the United Arab Emirates and Morocco have been the driving forces in the beginning, but today almost all countries are heavily involved in the energy transition. Solar and wind energy have become competitive. There is no more need for begging for subsidies. Renewables are swiftly becoming emancipated in the energy markets of the three neighbouring continents, Europe, Africa and Asia. Moreover, energy is getting exchanged from one area to the other depending on supply and demand. Local governments, industries and people, rather than foreign players, are becoming the prime winners. A convincing proof is the fact that Dii's main local partner, ACWA Power, has quickly become the number one in the region.

Today, in 2019, the authors observe a convincing case for renewables in all countries of MENA, even in politically unstable areas. As renewables in the deserts have become competitive across the board, the traditional subsidies for fossil energy have become superfluous, if not completely ridiculous. Governments are, courageously but still socially acceptable, gradually scaling down such subsidies. After years of wasting energy in the production countries and ever more costly import dependency in countries without oil or gas sources, a sort of 'normalisation' is taking place. Each desert country in the region has the same fair chances to make best use of indigenous solar, wind and perhaps hydro sources. The MENA region is quickly developing into a 'powerhouse' for emission-free electricity ('green electrons'), hydrogen and other energy carriers ('green molecules').

In fact, Dii evolved in ten years from Desertec 1.0 (power from the deserts for Europe) via Desertec 2.0 (emission-free power for MENA in the first place) to Desertec 3.0 (emission-free electrons and molecules for MENA and for the global energy market!

To date it almost seems hard to believe that there has been so much disbelief, ridiculing and confusion ten years ago. The process, from where Dii started then until today, has been fairly dynamic and sometimes bumpy, but it is a story of great progress and success. An exciting story, which will undoubtedly continue in the next decades. The German industry has embraced power from the deserts ten years ago as their love baby. That German love baby has now, so to say, become a young adult. It is no longer a remote 'Germany show'. The process is now driven by local stakeholders in cooperation with global partners. It has become unstoppable and will move on until energy supply in MENA will have become fully emission-free. Surplus energy will be fully available to the global energy markets. The regional and international industry will, hence, be ensured of many years of business opportunities in the regional energy transition.

> *Today, it may be hard to believe that in 2009, at the start of Dii, energy from the desert was perceived as something exotic, uncertain and very uneconomic.Today we know that the desert regions will swiftly become the powerhouses for emission-free energy.*

0.2 Establishing a 'not for profit' enabler for renewables

First, eleven mainly German companies and the Desertec Foundation (a German NGO) founded Dii (Desertec Industrial Initiative) in Munich, to discover during a period of three years whether the Desertec vision really has 'hands and feet' and how the idea could be implemented... In close coordination, a statute was laid down in advance which formulated the mission and goals. During the following three years, Dii would develop studies and ideas for framework conditions, some reference power plants and a roll-out plan for investments up to 2050. The project met with great public response. Expectations of a future that could depend significantly on energy supply from the desert were high.

Initially, most of the founding companies still adopted a wait-and-see attitude. At the board level, however, Desertec was a present topic: Attractive business opportunities were seen in a new market and an advance into the field of renewable energies seemed imaginable. The hope that the German government could extend the then generous promotion of renewable energies in Germany to the MENA region further increased the interest of companies.

The joint initiative of industry and the Desertec Foundation, which is still very young and committed to ideological goals, was welcomed by the German government and it pledged its support. However, there was no actual consultation with European stakeholders or governments, institutions and civil societies in the MENA countries before the creation of the initiative – the initiative was a real surprise.

Thomas Rüschen of Deutsche Bank acted as chairman of the shareholders' meeting in the early years. During this phase, the initiative grew into a unique international gathering of 20 voting companies and 35 associated partners. At the end of 2012, Frank Detlef Drake of RWE took over as Chairman. He guided the company through the high waves that hit a dramatic shift among the shareholders and associated partners. The Desertec Foundation withdrew from the initiative in mid-2013. By the end of 2014, Dii had completed the studies intended at the time of foundation and formulated investment ideas. Dii's key report 'Desert Power 2050' showed the long-term perspective for a (nearly) emission-free energy market in MENA and electricity exchange with Europe. In 2015, Dii relocated its operations to Dubai, an inspiring centre for innovation in the region capturing tremendous synergies. Desertec thus entered into a next phase, primarily concentrating on the MENA region. Since then, shareholders have been a Saudi company (ACWA Power), a German company (Innogy) and a Chinese company (CEPRI/SGCC). From the Arab metropolis, Dii, with the help of many international partners, is now working not only for emission-free power, but for emission-free energy in general in the MENA region and for the exchange of energy with the world market for emission-free energy, including Europe.

0.3 *Collegiality and Controversy*

The governance of a very heterogeneous industrial group is a dynamic process that requires special leadership skills from the Dii manager Paul van Son. In the beginning, the exchange with key persons in the shareholder circle took place quickly and unbureaucratically. In the background, very committed board members of the participating companies, such as Thorsten Jeworrek (Munich Re), Caio Koch-Weser (Deutsche Bank), Udo Ungeheuer (Schott) and Peter Smits (ABB), provided advice and support. The former Environment Minister and commissioner of the United Nations Environment Programme (UNEP), Klaus Töpfer, was available to the Dii leadership for some time as a special envoy. An Advisory Board, headed by Hans Müller-Steinhagen, director of the German Aerospace Center (DLR) and later dean of Dresden Technical University, also advised the Executive Board, which included the Tunisian State Secretary Abdelaziz Rassaa and the CEO of Royal Air Maroc, Driss Benhima. In addition, there were several external key advisors such as Gerhard Hofmann, former political chief correspondent of RTL and N-TV, Wolfgang von Geldern, former State Secretary of the federal government, and the law firm Hengeler Mueller from Munich. The relationship with the Desertec Foundation was ensured through its Founder Gerhard Knies member of the Club of Rome and Friedrich Führ.

The CEO of Dii was given great freedom to develop the company's programme and build a competent team. The relationship between the Management Board and the shareholders was characterised by trusting and collegial interactions. Compliance was ensured by explicit regulations and constant monitoring by the corporate group. Under Van Son, Dii used the Dutch 'polder model' in the coordination between the various companies with their different business objects, as the *Financial Times Deutschland* reported under the heading "Sent to the Desert" (Gassmann

30.10.2009). The cards were on the table from the beginning. All those involved worked together to find solutions to the problems, dilemmas and challenges that were plentiful in the first two years.

0.4 From electricity for Europe to energy structures for the region

However, this has not hindered the development of numerous collaborations. Together with the Fraunhofer Institute for Systems and Innovation Research, ground-breaking studies were conducted and intensive discussions were held with stakeholders from politics, institutions, the media and civil society in Germany, Europe, the MENA countries as well as China, Japan and the USA. The dilemmas of the Desertec idea and the realities of the market gradually became visible. The original, strongly constricted idea of building solar power plants in the deserts and bringing the electricity generated there to Europe proved unrealistic in this simple form. It was much more sensible to work with local governments to build a market in which renewable energies can compete. To this end, however, the necessary infrastructure and a fundamental openness of the market to the international and intercontinental exchange of energy expected in the long term must be guaranteed.

The idea of generating energy in the deserts of this world has long since lost the appearance of the exotic; in many countries it has become part of government plans. How and when an almost completely emission-free energy supply can be realized is the subject of countless controversies. But the direction is clear and the process is unstoppable. Many players are already active in the various local markets. Dii is still one of the driving forces behind this development, without receiving much media attention. It is a signpost with a wide network and access to decisionmakers in the MENA region and beyond. German companies have lost their pioneering role within the project. Arab, Asian and a few European companies are now taking the initiative.

1. Desertec – an idea gets outlines

The sun-drenched deserts of the MENA region are increasingly becoming the site of a massive energy revolution. The sun shines particularly intensely on the desert soil. This makes these desert regions one of the best places in the world for solar energy. Constant winds blow along the coasts. In the Gulf of Suez and in the Strait of Gibraltar and many other places, large wind farms have been built. As far as renewable energies are concerned, the early considerations on this matter began around 1990 in the context of the Club of Rome. Initially, it was politicians, scientists and economists from the Club of Rome and the Jordanian Energy Research Centre who developed a vision initially called Desertec.

The Sahara dominates the north of Africa. It stretches 6,000 kilometres from the Atlantic Ocean in the west to the Red Sea, which separates Africa from the Arabian Peninsula; between its northern and southern borders lie 2,000 kilometres. It is the largest desert in the world and extends over more than nine million square kilometres. The collective imagination connects the Sahara with gigantic sandy areas. In fact, only a quarter of the area is covered with sand. The majority consists of mountains, stone and gravel areas. The desert of deserts is joined to the east by other deserts: the Syrian desert, the Nefud desert and the Rub-al-Chali desert.

In these regions less than 50 millimetres of rain fall annually, which corresponds to one tenth of the Central European average. On the other hand, 700 times more solar energy reaches the surface than mankind currently obtains from fossil fuels (Martin 2016). The first considerations as to how this almost inexhaustible source of energy could be harnessed were aimed at producing clean energy using solar thermal power plants and distributing it to the world's population with low losses by means of high-voltage direct current (HVDC) transmission – more than 90 percent of humanity was to be reached in this way. Prince Hassan bin Talal of Jordan, the former president of the Club of Rome, suggested this model. The idealistic and visionary image was questioned, examined and improved in the following decades, because Desertec also turned out to be an extensive international search process in which the most diverse organisations and players were involved.

Figure 1: The North African desert landscape. Source: Thomas Isenburg

While very few people live in the deserts of the region itself, the coasts are densely populated. People also settled in fertile regions such as the Nile Valley. In 2018, North Africa had a total of about 200 million inhabitants. At the same time there were 500 million people in Europe. Unlike in Europe, however, the North African population is growing rapidly despite crises and challenges. The situation in the Middle East with its 350 million inhabitants is similar to that in North Africa. So, there are more people living in the MENA region, including Turkey, than in the European Union. This fact has interested scientists in Europe for some time, as well as energy and water issues in the region.

In 1968, long before the idea for Desertec was born, the Club of Rome was founded. At first it was a rather elitist discussion circle: its members were spread all over the world, preferably male and all in influential positions. However, when the study *The Limits to Growth* (Meadows et al. 1972) appeared in 1972, it initiated a broad debate on the environmental implications of our actions that extended far beyond the ivory towers. The book also had an enormous impact against the background of the first oil crisis in 1973, when the authors were awarded the Peace Prize of the German Book Trade. At the Massachusetts Institute of Technology (MIT), young researchers led by the American economist Dennis Meadows and his co-authors Donella H. Meadows and Jorgen Randers used computer-aided simulations to model the Earth's system behaviour over the next 130 years. Five different scenarios resulted: If world population, industrialization, pollution, food production and the exploitation of natural resources were to continue to develop as before, the world's absolute growth limits would be reached over the next hundred years. However, the book does not read exclusively pessimistic: its authors were convinced that the self-destruction of human civilization could be stopped by technical innovations and targeted control.

In the mid-1970s, the German physicist and philosopher Carl Friedrich von Weizsäcker explained the current state of knowledge on climate change to Federal Chancellor Helmut Schmidt. However, the first findings about this phenomenon go back much further. As early as 1824, Jean-Baptiste Fourier described how trace gases in the atmosphere can contribute to warming the climate. In 1860, the physicist John Tyndall proved that these are mainly water vapour and carbon dioxide (CO_2). In 1896, the

Swedish Nobel Prize winner Svante Arrhenius first predicted that doubling the CO_2 content of the atmosphere could lead to a global temperature increase of 4 to 6 degrees Celsius. In the 1930s, the experts then discussed an increase in carbon dioxide concentration as a result of industrialisation, which at that time was already associated with a warming of the earth's climate – a thesis that was, however, rejected due to the lack of reliable data. Since the 1950s, however, the danger of anthropogenic, i.e. man-made, climate change has been taken seriously in academic circles. During the international geophysical year 1957/58 it was proven that the CO_2 concentration in the atmosphere actually increased. Isotope analyses also showed that the increase was caused by carbon from the use of fossil fuels – i.e. by humans, as Stefan Rahmstorf and Hans Joachim Schellnhuber of the Potsdam Institute for Climate Impact Research explain in their book *Climate Change* (Rahmstorf/Schellnhuber 2007).

A driving factor for such search processes was increasing knowledge about climate change. One consequence was that in 1988 the Intergovernmental Panel on Climate Change (IPCC) established itself in order to bring scientific findings into the political and social debate. The IPCC's first progress report was published in 1990 and others were to follow. In view of the increasingly obvious consequences of climate change and the increasingly irrefutable responsibility of mankind for it, the Kyoto Protocol to reduce greenhouse gases was adopted in 1997. It sets the ratifying states the goal of limiting global warming caused by greenhouse gases to a maximum of 2 degrees Celsius. This means that the concentration of carbon dioxide in the atmosphere must not exceed 450 parts per million (ppm). At one of the follow-up conferences, COP21 2015 in Paris, this target was corrected to a temperature rise of 1.5 degrees. The burning of fossil resources must be reduced rapidly and drastically, and renewable energies must be massively expanded.

1.1 Gerhard Knies' vision

The pioneer of the initially different Desertec idea was the Club of Rome member Gerhard Knies. As a scientist at the Deutsches Elektronen-Synchrotron (DESY) in Hamburg, he studied the properties of matter in light – basic physical research. The Chernobyl Super-GAU of 1986 changed its perspective: the previous proponent of nuclear energy now increasingly considered its negative aspects, such as the problem of the production of nuclear weapons. So, he was looking for alternatives. In 1995, the physicist recognized the enormous potential of the deserts for the production of renewable, emission-free electricity – the idea of desert electricity has been with him ever since. Only six hours of solar radiation in the deserts of this earth would be enough to satisfy the hunger for energy of mankind of an entire year, was one of his fundamental considerations.

Figure 2: Gerhard Knies. Source: Thomas Isenburg

Knies, who died in 2017, was a rational man who, like many physicists of his time, included the social and ecological effects of his science in his considerations. Especially in nuclear physics, the philosophical element of scientific thought has a long tradition. Knies was driven by an interest in visions with expansive social and ecological dimensions. His core idea that the sun provided more than enough energy for the needs of mankind was presented in the mid-1990s as part of the scientific initiative 'Responsibility for Peace'. The idea was soon widely discussed, not only in view of the growing awareness of climate change and its consequences, but also as a potential contribution to the political and economic stabilization of conflict-ridden areas such as North Africa and the Middle East. Desertec could link the desert regions of the earth with the metropolitan areas. Large solar power plants would produce electricity from renewable sources in the sunny desert countries and this would flow into the European metropolitan areas through point-to-point connections, for example through the Mediterranean; in return, the electricity producing countries would receive financial resources for their development. For this principle Gerhard Knies created the name Desertec – short for Desert Technology – which was to have a great public impact.

Figure 3: A possible infrastructure for the sustainable supply of Europe, the Middle East and North Africa. Source: Dii

Initially, considerations and public discussion focused primarily on high-performance solar power plants with heat storage and basic load options. The first precursors of such power plants had already been built about 100 years ago in Egypt. Solar thermal power plants (Concentrated Solar Power, CSP) bundle solar rays through mirrors to convert the radiant energy into electrical energy via thermal energy. Its central property is the ability to store heat and make it available later, in the evening or during the night.

Most CSP power plants operate with sun-tracking mirrors in the form of a parabolic trough that direct the light onto an absorber tube in the focal line. The heat is stored at temperatures around 550 degrees Celsius using a heat transfer medium, for example in the form of liquid salt. Alternatively, solar radiation can be focused on the top of a tower and converted into thermal energy (FIZ Karlsruhe GmbH 2013). Steam turbines and alternating current generators, such as those produced by Siemens for over 100 years, generate electricity from this thermal energy.

Figure 4: A parabolic trough power plant in the model. Source: Thomas Isenburg

1.2 Prominent forerunners

Major renewable energy projects in Africa have a long tradition of electrifying the continent. One of probably the largest and boldest projects with considerable geostrategic conflict potential was the Aswan Dam in Egypt (Isenburg 2019). This was then actually realized.

For Egypt, the Nile has been of considerable importance for thousands of years. Its changing water levels and flooding with sediments rich in nutrients for agriculture give the fields great fertility. During periods of drought it can also lead to considerable crop failures. These fluctuations could be avoided by regulating the water level. This could be achieved by a huge reservoir with an impressive dam wall. In addition, a hydroelectric power plant would be able to provide the country with electricity, according to initial considerations about 100 years ago.

At first it was British engineers and scientists who took on the gigantic task. It was the engineers William Willcocks, Benjamin Backer, William Edmund Garstin and the mathematician Harold Edwin Hurst, who took care of the construction of the dam and the effects of the dam of the enormous masses of water.

On the Arab side, it was Adrian Daninos, a Greek-Egyptian agricultural engineer, who was involved in the construction of the dam as early as 1912. With the inheritance of his father in the amount of 100,000 pounds that he received in 1925 he financed the development of the project. With his commitment he wanted to change the standard of living and the fate of the Egyptian people.

At the end of the 1940s he first contacted King Faruq, who was not interested. Four days after the fall of Faruq, caused by Nasser, the plans were seriously discussed by the new rulers, as Daninos contacted officers close to Nasser. Forgotten and impoverished, Adrian Daninos died of the consequences of an accident at the age of 89 in 1976.

When Nasser took power, the search for implementation options for the enormous project began. It was to be a building project that was unique for the world and was noticed by the superpowers USA and Soviet Union.

At first, German companies had a good chance of winning the contract. In the mid-1950s, the German Hoch/Tief und Rheinstahlunion made an offer for the construction of bridges. The companies wanted to build a rockfill dam and a power plant with 8 turbines. However, financing for the costs estimated at 2.1 billion euros was still lacking. Nasser's idea was to finance the increased costs through the proceeds of higher rice harvests. However, the cold war put a stop to this calculation, and Egypt acknowledged the People's Republic of China. The USA and Great Britain withdrew and the Soviet Union jumped into the breach.

In particular the American Secretary of State Dulles in the Eisenhower government did not trust Nasser. He believed the project would collapse without American help. However, the Soviet Union jumped into the breach and supported the country on the Nile. In 1956, French and Israeli troops attacked Egypt on the pretext of protecting the Suez Canal. This could have caused a confrontation between the US and the Soviet Union. Eisenhower joined the strict refusal to support the invasion of Britain and France. The invasion forces withdrew from Egypt at the beginning of 1957. US relations with the Middle East were damaged. A conflict between the great powers had to be avoided. Eisenhower did not support France and Great Britain. They withdrew again, but the relationship between the USA and the Middle East was busy. In 1964 the Aswan dam was put into operation. The reservoir is about 500 kilometres long. The area data reaches up to 6,000 square kilometres.

Figure: 5 Aswan Dam in Egypt Source: voith

The central construction element is a 3,800-meter-long dam wall. It is 111 meters high. At the bottom it has a width of 980 meters and at the top 40 meters. The dimensions alone show the efforts that were necessary to erect the enormous structure. In addition, there was a hydroelectric power station with 12 Francis turbines. They generate an output of 2,100 MW. The electricity generated is routed to Cairo. At the start of operations, the power plant supplied 50 percent of Egypt's electricity. The consequences of longer periods of drought for agriculture have also been reduced. In this way, the Nile valley could be protected from flood damage. However, fish stocks have been seriously damaged.

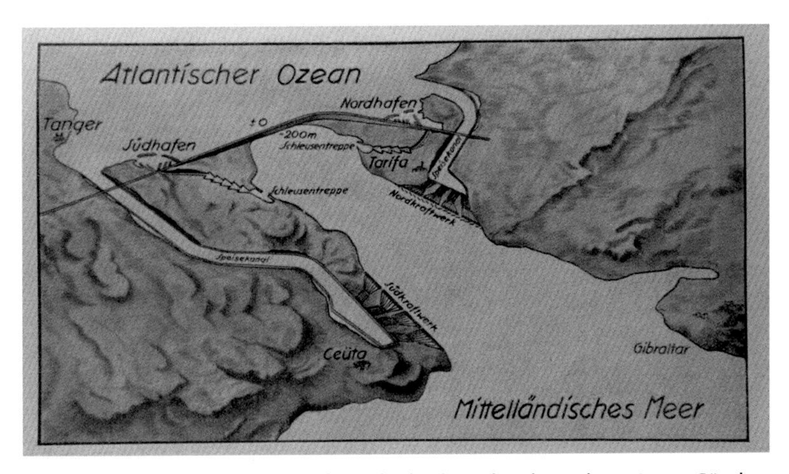

Figure: 6 The Atlantropa project from the book: In hundert Jahren,Hanns Günther. Puplished in 1931.

The project 'Atlantropa' is regarded as a precursor of the Desertec version. It was run by the German architect Hermann Sörgel since the twenties. (Günther 1931) Here as well, the main work would have been done by the sun. The architect Sörgel wanted to transform the Mediterranean into a huge inland sea. Atlantropa wanted to follow on from the great engineering achievements of the 19th and 20th centuries, such as the connection between the east and west coasts of the USA by rail, the Suez and Panama Canals and the gates with which the Dutch closed off the sea. Sörgel calculated it this way: The Mediterranean Sea loses 3850 cubic kilometres of water annually through evaporation. These are offset by 1250 cubic kilometres of precipitation and 180 cubic metres of inflow from rivers and the Black Sea. The missing volume flows through the Strait of Gibraltar from the Atlantic Ocean, according to analyses of the time. The huge dams on the Straits of Gibraltar and the Dardanelles were to lower the sea level of the Atlantic Ocean by 200 metres. This could result in a land gain of around 500,000 square kilometres. In addition, the dams were intended to produce electricity. Ecologically, the project would have had persistent consequences. Geopolitically, however, Atlantropa would have replaced Europe's east-west orientation with a north-south concept that combines Europe's capacities with Africa's resources.

The foundations of European thought had developed around the Mediterranean – a special connection between familiar and foreign, between mutual knowledge of what lay on the other side of the sea.

For the Gibraltar dam, Sörgel wanted to remove parts of southern Spain and at the same time develop Africa. Until the 1950s the project was very popular. Then it was discarded.

1.3 *Good climate for sustainability*

At the turn of the millennium, the Desertec vision was again characterized by a favourable climate. In December 1997, the EU ratified the Kyoto Protocol on behalf of the 15 countries that were members of the European Union at the time the agreement was signed, thus shaping the United Nations Framework Convention on Climate Change. In many European core countries, social democratic governments were in charge; a comprehensive trend towards more sustainability was also supported by broad currents within parliaments. In Germany, for example, in 2000 the red-green federal government, which had been in power since 1998, passed the Renewable Energy Sources Act, which was intended to accelerate the expansion of renewable energies; its spiritual fathers were 'der Grune' Hans-Josef Fell and the social democrat Hermann Scheer. Michael Eckhart, a banker with Citigroup, convinced Hermann Scheer of the need to safeguard investments in the long term by protecting feed-in tariffs. In this sense, the law guaranteed a fixed feed-in tariff for electricity from renewable energy sources – a regulation that would lead to numerous controversies. Somewhat later, in 2007, Spain also created a subsidy for electricity from renewable energies, but renounced state protection.

The Netherlands and the Scandinavian countries, on the other hand, opted for a different concept: they had already had good experience for several years with the state-certified trading of green certificates (Renewable Energy Certificate System, RECS), based on Guarantees of Origin (GoO). However, the EU did not support RECS's well-monitored international trade by renewable energy producers, not least because Germany and Spain put pressure on the European Council to adopt new renewable energy guidelines in 2008. Germany feared that certificate trading could undermine the promotion of renewables through feed-in tariffs, as in the medium term it would channel investments to the countries with the most favourable conditions for renewables – the German representatives saw this as a danger for investments in renewables in Germany. This, however, prevented the government from creating transnational and intercontinental opportunities for action. The green certificate instrument would have enabled consumers across Europe to opt directly or indirectly for green electricity from North Africa, thus promoting new installations in the Mediterranean. The adopted EU directive, on the other hand, only provided for a complicated, market-agnostic possibility of subsidising installations outside the EU that were electrically connected to the EU network. This regulation has not yet acquired any practical relevance.

Nevertheless, the Desertec project attracted a lot of attention and around the project a network of players from science, economy and politics was formed. On the initiative of the Club of Rome, Gerhard Knies founded the Trans-Mediterranean Renewable Energy Cooperation (TREC) in 2003. The drivers behind international networking were the Hamburg Climate Protection Fund and the Jordan National Energy Research Centre (NERC).

It was then in Amman, Jordan, that the TREC announced its perspective in October 2003: the constantly growing world population, the gigantic differences between rich and poor and the ever-increasing demand for energy virtually dictated the establishment of a market for renewable energies linking the Mediterranean countries. To tackle this ambitious project, experts from Egypt, Benin, Germany, Jordan and Morocco were invited by the Club of Rome to draw up a master plan.

1.4 Studies, studies, studies

DLR attempted to identify the first well-founded scientific impulses in Germany. The Federal Ministry for the Environment, Nature Conservation and Nuclear Safety (BMUB) provided DLR with the funds for this purpose. The project leader was Franz Trieb, an expert in the storage of renewable energies. Now he was to sound out the opportunities for huge solar thermal power plants for the Mediterranean region and put Knies' visions on a scientific basis. In addition to German scientists, the team also included researchers from Morocco, Egypt and Algeria.

The first of three DLR studies on the feasibility of the desert energy project was published in March 2005 (DLR 2005). The research groups examined possible scenarios for supplying the countries of Southern Europe, North Africa and the Middle East, known as EU-MENA states, with wind and desert sun. The focus was not only on the technical dimension, but also on the question of social sustainability. With the study, DLR pursued the goal of providing decisionmakers in politics and business with background information on local development options. It stated that the establishment of appropriate market introduction instruments was necessary for a fruitful strategic partnership between the EU and the MENA countries. In principle, these could be based on the feed-in tariffs established in Germany and Spain for electricity from renewable sources.

Large parts of this first study dealt with the analysis of the resources available in North Africa and the Middle East for power generation from biomass, solar energy, wind and hydropower as well as geothermal energy. DLR concluded that there was an abundance of renewable energy sources in the EU-MENA region and that they could easily meet the growing energy needs of these countries with their growing populations; in the medium term even an export of renewable energy to Central Europe was conceivable. Solar radiation was identified as by far the largest available resource: its total potential in this region alone exceeds world electricity consumption many times over.

According to the DLR study, the years between 2010 and 2025 would see a massive expansion of large solar power plants with outputs of up to 2,000 megawatts each. The authors forecast a drop in prices, especially for photovoltaic systems, but also for large solar thermal power plants. In the same period, however, the consumption of a country such as Morocco, which at the time of publication of the study was covering its energy needs from purchased fossil raw materials and electricity supplies from Spain, would also increase considerably. Population growth in the MENA region has been identified as the driving force behind growing energy consumption, which could rise to 3,500 terawatt hours per year by 2050 – the equivalent of Europe's total demand in 2005.

Although the focus of the study was on solar thermal power generation, other renewable energies were also considered. The 2005 state of the art result was based on the assumption that fluctuating sources such as hydropower, wind power and photovoltaics generate relevant quantities of electrical energy but can only make a relatively small contribution to the guaranteed output. Therefore, the basic study preferred more controllable sources such as biomass (whereby the authors emphasized that their use would in no way compete with food production), geothermal energy and storage water, and in particular solar thermal power plants.

The study represented a first reliable result for the clarification process that is now beginning with regard to the feasibility of the Desertec vision. Early results were inconclusive, they had to be constantly checked and improved – this is typical for scientific knowledge processes and was no different in this case either. When the study was published, the renewable energy markets in the MENA region were not yet developed. Only well-known hydropower plants, such as the Aswan Dam in Egypt, were in operation and a handful of wind farms were being built. The region was still almost entirely dominated by fossil fuels. An environmental awareness, as it had long been widespread in Europe, first appeared slowly in the Arab world and initially only among the elite.

The second DLR study entitled *Trans-Mediterraner Solarstromverbund* (Trans-Mediterranean Solar Power Interconnection) (DLR 2006), which begins with a foreword by Jordanian Prince Hassan bin Talal, was published at the Hanover Fair 2006 in Germany. From 1999 to 2006, King Hussein's brother was President of the Club of Rome. He is an extremely polite, almost quiet man, who is particularly interested in intercultural communication and who wants to bring Europe, the Middle East and North Africa closer together. In the foreword he writes: "The countries of the sun belt and the technology belt of the earth can become very powerful if they recognize themselves as a community: a community for the security of their energy and water resources and for the protection of the earth's climate; a community for their common future." (DLR 2006)

The event at the Hanover Fair was intended to prepare a top-level dialogue between entrepreneurs, politicians and energy experts from all over the world. Technological as well as economic and ecological trends in worldwide energy systems were discussed. In this context, DLR's second study attempted to provide the first concrete results on aspects such as economic efficiency, compatibility, environment, society and security of supply for the gigantic project, whose capital requirement was estimated at 400 billion euros. The paper concluded that a balanced mix of renewable resources and fossil control capacities would allow a secure supply of electricity in line with demand, without placing an unacceptable burden on fossil energy sources and the environment. The authors have gone as far as to predict that a ring of HVDC lines around the Mediterranean could be closed by 2010.

Further connections between the continents were to run through the Mediterranean Sea as submarine cables. It was also planned to connect the continents via Jordan and Turkey by high-voltage lines. According to the scenario, by 2050 around 700 terawatt hours of solar energy could be fed into a European supergrid annually from the MENA region via 20 intercontinental cable and grid connections. The MENA electricity could even be routed as far as Scandinavia, Great Britain and Iceland by means of such connections.

The study's forecast was based on the situation in 2000, when renewable energies already accounted for 20 percent of the European energy mix, mainly due to the expansion of hydroelectric power plants. This would rise to 80 percent by 2050. According to the optimistic assumptions of the publication, the negative consequences of fossil energy supply would already be overcome by 2020. By 2025, 60 terawatt hours of renewable energy from the MENA region could already be available, and by 2050 the 700 terawatt hours mentioned above would already be available. In connection with the low-loss HVDC lines, electricity costs of around 5 eurocents per kilowatt-hour would be incurred, while carbon dioxide emissions would be reduced by 25 percent. Energy partnerships or free trade areas would be established between the countries of the EU-MENA region. In addition, the study assumed price reductions for solar power plants and the development of an efficient HVDC technology as well as politically stable relations in the MENA region with a further easing of world politics.

DLR's work was ground-breaking for its time. However, the emphasis was almost exclusively on solar thermal CSP power plants in the deserts, which were to make a major contribution to the standard output in Europe, including Germany. Energy experts such as Detlef Drake from RWE in Essen and Hervé Touati from E.ON in Düsseldorf turned their attention to new technology at that time with power plants in a centrally organised form, as this fitted in well with the thinking of the energy industry at the time. It was hoped that rapid progress would be made in technological development as a result of which electricity prices would become marketable. Photovoltaics, on the other hand, was still regarded with scepticism, due to the high costs and the interruption of production during hours of darkness. RWE CEO Jürgen Großmann, who was in office at the time, told the press: "Photovoltaics in Germany is like growing pineapples in Alaska" (Großmann 2012). In addition to dynamic technological progress, the company relied on a developing power grid in a sensitive region without any practical intercontinental market experience. The path of the project seemed to be mapped out against this background.

Based on DLR studies, the Club of Rome formulated a White Paper entitled *Clean Power from Deserts – The DESERTEC Concept for Energy Water and Chemicals and Climate Security* (German Society Club of Rome 2009). In November 2008 Prince Hassan of Jordan and Gerhard Knies handed it over to the President of the European Parliament, Hans-Gert Pöttering.

1.5 From vision to reality

Initial project estimates for the Desertec project indicated an investment volume of 400 billion euros for power plant outputs of 10,000 MW. These dimensions caused confusion and criticism from the very beginning. Representatives of the CSP industry pointed out that they would already be happy if they reached 50 megawatts with the proposed solar thermal technology – this output would now be increased by a factor of 200. Knies had calculated that the ten billion people on earth expected in 2050 would require around 60,000 terawatt hours of energy per year. About half of this requirement was to be covered by solar power from the desert, with Knies taking the 3,000 hours of solar radiation in the regions near the equator as a basis.

With the interest of politics, the interest of the economy also grew and Desertec began to stimulate the financial world. The Münchner Rück, now known as Munich Re, began to take an interest in the project at an early stage. Large insurers in particular are affected by the considerable losses caused by extreme weather conditions as a result of climate change. Peter Höppe was Munich Re's head of the Geo Risks Research / Corporate Climate Centre department at the time. Renewables were regarded as an important growth area.

In May 2009, a founding workshop was held to prepare the ground for the Desertec Industrial Initiative. The participants were: ABB, Cevital, Deutsche Bank, E.ON, HSH Nordbank, MAN, Ferrostaal, Munich Re, Schott Solar and Siemens, as well as German Federal Foreign Office. Of the later shareholders, only Abengoa Solar, MAN, Solar Millennium, M&W Zander and RWE were missing – an alliance that had never been known in this form before.

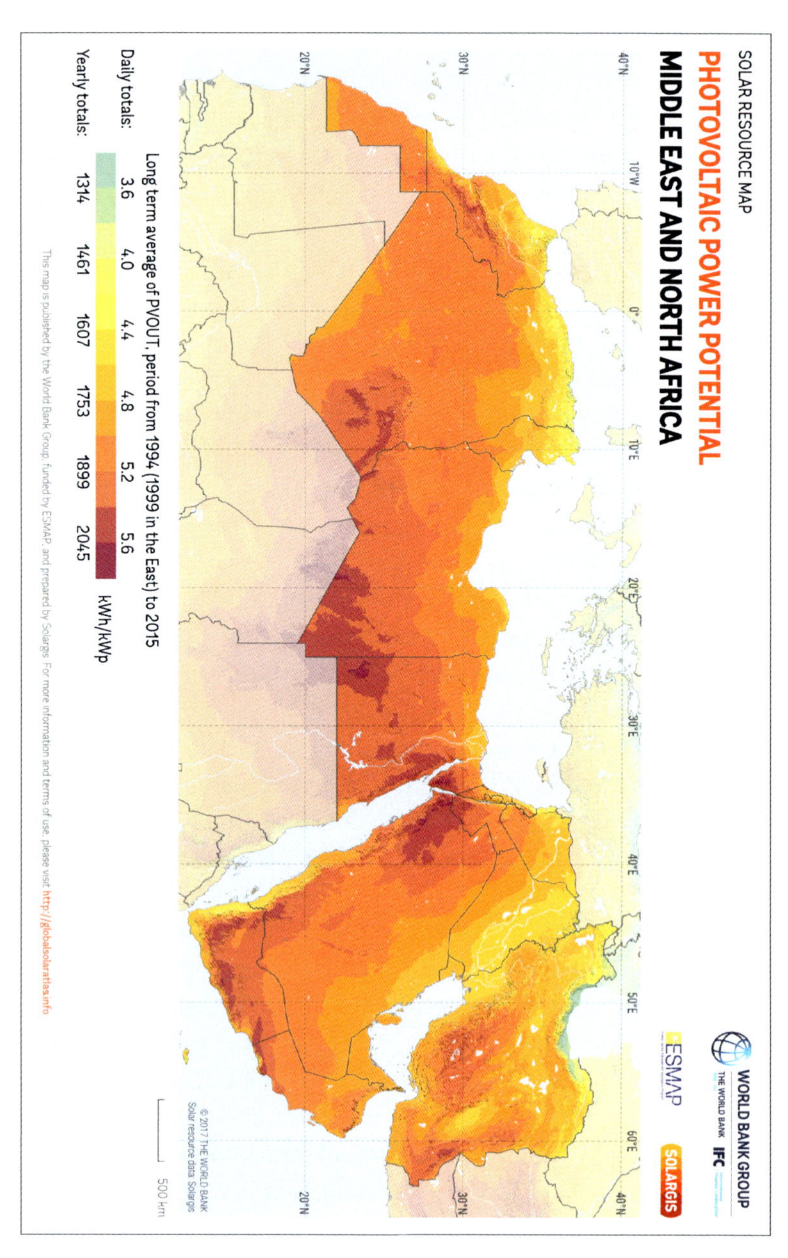

Figure 7: Sonneneinstrahlung: Source / Copyright The World Bank, Solarressourcen: Solargis

An interview with the *Süddeutsche Zeitung* in June 2009 by Torsten Jeworrek, who was constantly on the move during this phase to recruit partners for the Desertec vision, caused a sensation. Jeworrek mentioned Desertec only marginally as a possible part of his company's sustainability strategy. But he had aroused the curiosity of the SZ editors: 400 billion euros for electricity from the desert – the superlative became the lead story: "Desert electricity for Germany – Solar electricity from the Black Continent: A consortium of companies wants to build solar power plants in Africa – for 400 billion euros. It is one of the biggest private green power initiatives of all time." (Balser/Fromm 2009)

The article acted like an initial spark. Dozens of newspapers, specialist publications and online media now reported on the sensational project, bringing their own reflections, interviews and documentaries. Work continued behind the scenes – and in July the public was informed about a Memorandum of Understanding and the intention to found Dii.

Jochen Kreusel, professor of electrical engineering and proven expert for power grids, represented ABB from the German city of Mannheim in the consortium and remembers that the article in the *Süddeutsche Zeitung* had significantly increased the pressure to act. He reports that otherwise his company would probably not have entered into a partnership with its competitor Siemens so quickly. Others thought similarly, everyone wanted to be there.

The press conference acted as a powerful kick-off. The negotiating partners had also met regularly before but were now invited to weekly meetings. Munich Re continued to be a key driver of the project, seconding staff to the project. This phase, from April to October 2009, the participants experienced as highly exciting, but not only that: during this time a real relationship of trust grew between the contributors.

Finally, on 30 October 2009, the shareholders' agreement for Dii GmbH was concluded and a managing director was appointed. The representatives of the shareholders, who had come to Munich, the company's headquarters, on this day agreed: This was a historic moment with great impact.

1.6 Criticism of the 'Project'

Nevertheless, critical voices also spoke out on the plan, which was often referred to as the 'Project'. Fritz Vahrenholt, head of RWE's Innogy eco-electricity division, said: "What I don't like about the whole discussion is that people believe we're simply building a pipeline to Germany and then we have desert electricity (Manager Magazin 2009). Instead, Egypt, Tunisia or Morocco would have to benefit from the energy first – until the first electricity from the Sahara could arrive in Germany, there is still a long way to go. In addition, Vahrenholt called for state start-up financing along the lines of the Renewable Energy Sources Act.

One of the most prominent critics of the project was Hermann Scheer. As a member of the federal executive of the SPD in Germany, he had a decisive influence on the party's environmental and energy policy. He was regarded as a lateral thinker who was respected by all sides for his forward-looking visions. Hermann Scheer, who died in 2010, was awarded the Alternative Nobel Prize in 1999 for his commitment to social and ecological sustainability.

In his book *Der energetische Imperativ* (Scheer 2010), published in 2010, Scheer continued his criticism, dedicating an entire chapter to Desertec. He points out that the Desertec project, with a volume of 400 billion euros, was the largest investment project in global economic history. More than 40 countries would be involved, making the number of institutions involved unmanageable – the project was overly complex, Scheer said. In Scheer's eyes, Dii performed a daring balancing act by bringing together antipodes such as the industrial groups RWE, E.ON and Siemens, which had previously earned their money with nuclear power, and NGOs such as the Club of Rome and Greenpeace.

He justified the initial euphoria that Desertec triggered and the broad support it received in Germany and parts of the EU with the fact that a flagship project had finally been found in which ecology and the economy seemed to be combined and behind which everyone could gather. However, Scheer opted for decentralised solutions, because he feared that the large energy companies would exercise too much control – at a time when the decision to phase out nuclear power following the Fukushima nuclear accident had not yet been taken, and when an extension of the operating lives of the nuclear power plants was on the agenda of the German federal government. He justified the fact that many voices in the media and in the general public nevertheless placed great hopes in Desertec as well as in offshore projects in the North Sea with the technical feasibility: The storage deficit in solar power in Germany would be overcome because the planned solar power plants could produce electricity throughout. In this respect, he rated Desertec as captivating.

However, he describes the project as too strongly focused on Germany, both with regard to the composition of the consortium and the energy industry interests articulated there. In Scheer's eyes the project was designed purely for the energy industry, not for the economy as a whole and certainly not for the regional economy. It reduces the number of renewable energy players instead of increasing them. Scheer was also critical of a transmediterranean supergrid: If this is not achieved quickly, energy companies could be justified in continuing to operate their large power plants. This is an à la carte concept for the power companies, as they gain time during the long construction phase to continue running their profitable power plants based on fossil resources. The supergrid would have to be completely redesigned and would extend from the desert regions to northern Europe – a gigantic scale for which there was no model. The longest HVDC power lines to date stretched over a length of 2,000 kilometres in China; at Desertec, however, these lines were to be up to 5,000 kilometres long. In addition, only point-to-point connections had been made before, never a complex network such as Dii wanted to implement. Experience with HVDC cables at a depth of more than 1,000 meters was also lacking.

In 2018, almost ten years after Hermann Scheer and other fundamental critics had voiced their criticism, we can see that many objections to the Desertec project, which was still rather rudimentary, were justified. As this book will show, the emphasis on huge solar thermal power plants, the fastest possible export to Europe and the need for an all-encompassing supergrid reflected a lack of understanding of the complexities and sensitivities in the regions and markets involved. Scheer's fear that visions, ideas and plans from the laboratories and heads of industrial captains would be implemented one-on-one was not realistic. Although the Dii movement was initiated by large companies, it was embedded in a broader context with disparate stakeholders. At the time, German industry was actually not even prepared to develop and implement a 400-billion-euro project. It was particularly interested in finding out whether the deserts were suitable for securing an emission-free future energy supply and whether profitable projects could be realised there. Scheer may have been right about the fact that the top dog mentality prevailing in the energy industry at that time would prevent rapid changes – nevertheless all companies involved loyally supported the Dii studies. Dii has proven to be an important guide in this environment, and the process has proven to be transparent and precise. In the next chapters we will examine the process from the German-centric Desertec idea to the now rapidly developing renewable energy markets in the MENA region – an exciting if not exactly straightforward process.

2. The starts of Dii (2009-2011)

2.1 *Dii learns to walk* (2009-2010)

The industrial consortium, initially dominated by Germany, was rated very positively in the press. Knies had achieved one of his goals: his vision had generated publicity. This stimulated social dialogue – but also raised high expectations. The centre is initially in Germany, around the location of Dii in Munich. In different countries there are also individual people who support the Desertec idea. Examples can be found in France (Francis Petitjean), UK (Gary Wolf), Luxembourg (Oliver Steinmetz), Netherlands (Paul Metz) and Tunisia (Mouldi Miled).

After the spectacular performances of the plans at the kick-off in 2009, many internationally active companies wanted to jump on the bandwagon. Among the first were Enel, Italy's largest electricity supplier, Red Eléctrica De España, Spain's grid operator, and EDF, France's second-largest electricity supplier. In addition, there were companies from Morocco, Tunisia and Egypt. Since the initial initiative originated in Germany, more and more companies that do not have their headquarters in the founding country of Dii are finding their way into the group of companies. Terna and Red Eléctrica De España, the grid operators in Italy and Spain, have a great deal of know-how when it comes to electricity grids. For legal reasons, German network operators are rather cautious here. The Desertec plans and the large influx of companies calls France to the plan and the French President Nicolas Sarkozy wonders: what are the Germans doing in my backyard? He means the Maghreb. As a former colonial power, the countries of Tunisia, Algeria and Morocco belong to the French sphere of influence.

Moreover, in July 2008 the 'Union for the Mediterranean' (UfM) was founded as an intergovernmental association of 43 states around the Mediterranean Sea. It aims to promote dialogue and cooperation in the Euro-Mediterranean region. (Dubessy 2010) The 28 EU Member States in 2010 and 15 Mediterranean countries from northern Africa, western Asia and southern Europe are part of this group. The organization had developed its own solar plan. This provides for the development of 20 GW of capacity in the form of renewable energies. To this end, 50 billion euros would be invested by 2020. In May 2010, the Transgreen consortium was founded in Cairo in the presence of Egypt's Energy Minister. This organisation can be attributed to the French President's environment and is a response to the initial German ambitions of the Desertec. Officially, the two organizations would complement each other. The organisation of Transgreen is similar to that of Dii, but the company has a different objective. (IPAMED 2010) The company is committed to accompanying the development of the electricity grid.To this end, it intends to link up with the Moroccan solar plan and the power lines under the Strait of Gibraltar between Morocco and Spain. Motivated by Knies' ideas, the Moroccan King Mohammed VI launched a plan to develop renewable energy capacity in the form of solar and wind power plants. The monarch wants to disconnect his country, which is poor in primary raw materials, from the expensive supplies of fossil raw materials and is therefore focusing early on renewable energies. Tunisia is also developing a plan to implement renewable energies, albeit on a smaller scale. Transgreen proposes the development of further electricity highways between Algeria and Spain, Tunisia and Italy, Libya and Italy, Egypt and Greece.

The team around Van Son in Munich initially devoted itself to the general conditions for the implementation of the Desertec version in the framework determined to date by DLR. Cooperation between Dii and Transgreen has not yet taken place at team level, but a friendly relationship is developing between Van Son and Transgreen director Andre Merlin. Both energy managers have known each other for many years. Meanwhile, the French have had their issues about the rights to the name Transgreen. That will soon be solved by a new name, Medgrid. Medgrid, just like Dii, succeeds in winning influential partners. These include Siemens, ABB, Abengoa, which participate in both Dii and Medgrid, as well as a number of more grid-oriented companies.

Abu Dhabi, pioneer in the East

In the eastern part of the MENA region, too, the focus is on renewable energies. The oil-rich Gulf States are gradually becoming aware here and there that their fossil resources are finite. In the Arab Emirates, the ecological city of Masdar City is planned for 2008. Masdar is the Arabic word for source. The eco-city would no longer generate carbon dioxide emissions. To this end, renewable sources must take over the energy supply. The International Renewable Energy Agency (IRENA) is to give the project an additional boost by relocating its headquarters to this eco-city location. IRENA, together with the International Energy Agency (IAE), is responsible for energy issues in the field of renewable energies. By 2018, more than 170 countries will be members of the organisation, including the EU. A solar-thermal parabolic trough power plant could take over part of the power supply, is the consideration in the Gulf region. The construction of the solar complex with the name Sham began in 2009 and when it was completed in 2013 it could boast an output of 100 MW. The power plant will be operated by Masdar with a 60 percent share as well as the French mineral oil company Total and the Dii shareholder Abengoa, who will share the remaining 40 percent. Masdar is a company based in the United Arab Emirates. It is regarded as a world leader in the fields of renewable energies and sustainable urban development. The power plant operated by the three partners was at the time considered to be the largest of its kind and, according to the operators, it could reduce carbon dioxide emissions by 175,000 tonnes. The same effect could have been achieved by planting 1.5 million trees or taking 15,000 cars off the road. Sham occupies an area of 2.5

square kilometres. This corresponds to 285 football pitches in the desert near Abu Dhabi.

The 'major project' of the Desertec Industry Initiative

Meanwhile, a lot of media in Germany are taking up the topic of Dii again and again. Often, however, the portrayal is extremely condensed, to a 400-billion-euro major project that solar thermal CSP power plants are building in North Africa to supply electricity to Europe. This one-sided approach in the German media and through politics influenced the public and many companies in Europe as well as in the MENA region and worldwide. Maybe that is why there were no initial questions about the specific role and tasks of Dii. The attractive vision of the future, in which Europe's energy supply would be guaranteed by African desert electricity, was widespread and was strongly fuelled by industry.

Quite a few suspected that Dii already had over 400 billion euros at its disposal and, like a construction company, would now invest in power plants and electricity grids in the MENA region. In addition, the broad planning horizon of the project – 40 years after all – was completely ignored in most observations, which conveyed a false and extremely exaggerated image to the German public. The central task of Dii was rather to create fair conditions for desert electricity, to open up long-term perspectives for a low-emission or zero-emission energy supply in the MENA region and to facilitate the technical and organisational export of electricity to Europe. The guiding principle was and still is to produce desert electricity cheaply and on a long-term basis where there is sufficient sparsely populated land available. The countries of the Middle East and North Africa are to be put in a position to sell electricity on a permanent basis directly via lines or indirectly, for example using hydrogen as an energy source to Europe or India.

The energy experts hoped that the connection of the electricity markets alone would generate major synergies. For this to become a reality, all those involved in wind and solar energy still had a lot to learn. Internationally active shareholders such as Munich Re and ABB called for the creation of a suitable power plant mix to be approached as openly as technologically possible. Other shareholders such as Schott Solar, Solar Millennium AG and Abengoa pushed for the construction of solar thermal power plants as quickly as possible – in order to get their projects up and running quickly. Successful commissioning of CSP power plants in the USA, Spain and the United Arab Emirates (UAE) fuelled this hope.

Government support was needed for the start-up, but in the long run a cost reduction through technical progress was expected. Together with the climatically favourable conditions for the production of renewable electricity in the MENA region, many managers thought in 2009 that renewable energy projects would be able to assert themselves in a fair market environment without subsidies. However, information about the massive subsidisation of fossil resources in the MENA region has penetrated again and again. According to a publication by the International Energy Agency (IEA) in 2011, the subsidy for fossil fuels amounted to about 523 billion US dollars per year – a real uphill battle for renewables (Uken 14.10.2013). Even if Dii did not seek competition with the fossil economy at all.

The public perception of Dii and the Desertec idea in the MENA region at that time was a completely different, even almost non-existent. The idea from the German thinktanks had not yet gained a foothold in the Arab world. Confusion and irritation prevailed, which came up cautiously in personal conversations, but not in the press. The only key person to make a statement in 2009 was the Algerian Energy Minister Chakib Khelil. To foreign journalists he stated that he objected to the export of solar energy from his country by foreign companies. Commitment to desert power by powerful individuals, such as Prince Hassan of Jordan or the Moroccan King Mohammed VI, and individual government initiatives (e.g. the Moroccan Solar and Wind Plan 2020) were the exception. Even the scientists at the universities in Casablanca, Tunis and Cairo were hardly aware of the Desertec idea. In order to transfer knowledge from the North to the South – and vice versa – Knies had therefore proposed the establishment of a Desertec University. European and Arab students would expand and exchange their knowledge about renewable energies, power grids and corresponding markets. Although this approach has not yet been pursued, but a Desertec University Network (DUN) was founded in 2011.

In Europe, the majority agreed that Desertec would be technically feasible as the DLR studies had already proposed in 2005. The long-distance transmission of energy under the Mediterranean seemed feasible, but difficult. The sea between the continents is in some places over 2,000 meters deep, with a jagged underground. Energy transport was to be provided by HVDC lines that transmit large quantities of energy with particularly low losses and have been operated between England, Europe and Scandinavia since 1980. An example in the Mediterranean is the Sacoi connection from Italy via Corsica to Sardinia.

Nevertheless, questions remained unanswered. How, for example, the power grids would have to be designed on both sides in order to transfer over 100 terawatts safely. Or whether the mirrors of CSP power plants could function under the harsh desert conditions, with extreme weather conditions and sandstorms. The industry consortium was expected to provide answers to these questions and also to the possible effects of terrorist attacks. Many still doubted the project. There was a justified fear of political instability in the MENA region. Likewise, France's colonial past is still very much present in the Maghreb, especially in Algeria. The profitability of the project in particular was frequently put to the test during these days. People still did not believe in the competitiveness of renewable energies.

Especially the German public is discussing it. The perceived trend towards neo-colonialism, the hegemony of large companies and greenwashing were also criticised. The question whether desert power can be a sensible solution for local energy supply in MENA was hardly raised. The MENA countries were only considered capable of financing the project on the condition that they had the possibility to transport electricity to Europe. Germans and Europeans rarely succeeded in adopting the perspective of the affected MENA inhabitants. At the same time, they were hardly aware of the enormous wealth they possessed in the form of renewable sources (Lubbadeh 1.12.2009).

2.1.1 *The first steps*

Dii first wanted to study how the energy markets of the MENA region and Europe were connected and what interactions were possible. To clarify this, it looked at the three continents of (North) Africa, (West) Asia and (South) Europe and examined which generation and infrastructure capacities could be developed. Funding opportunities were also evaluated, such as special conditions from development banks, Power Purchase Agreements (PPAs), special feed-in tariffs, Renewable Energy Certificates (RECs) and CO_2 certificates. The subjects of investigation were also the essential technology and the required transfer of knowledge.

According to DLR considerations and studies, by 2050 15 percent of European electricity consumption and a significant proportion of local consumption would be covered by renewable sources in the deserts. This approach was based on physical and geographical considerations. This was based on a higher goal, namely, to free the electricity supply in the region and its European neighbours from carbon dioxide emissions at the lowest cost and at the same time create new, local jobs.

The players were faced with the great challenge of integrating a technology mix of photovoltaic plants, wind and hydropower plants as well as solar thermal plants – and in the transition phase also fossil fuels and nuclear energy – into a huge power grid. Solutions to this problem were not yet in sight in 2009. Forecasts for the realities to be expected in 2050 did not stand up to critical scientific scrutiny. The industry could only guess what leaps in development and market shifts they would face.

This uncertain future and diverging business interests of the industry, also within Dii, led to tensions. On the one hand, there was disagreement as to whether to commit to one technology or to approach the project in a technology-open manner; on the other hand, there was too much focus on a fast, physical export of electricity from North Africa to Europe. Some participants pushed for a 'German-style feed-in tariff', as if Moroccan electricity were generated in Germany. In theory, this was already conceivable, based on the so-called Article 9 of the European Renewable Energy Directive (Directive 2009/28/EC).

Where renewable energies were not yet competitive, and this was the case in 2009 in the MENA region, support mechanisms had to be developed in international cooperation. The possible instruments were known: Initial funding by governments, public funding by institutions and foundations, development banks, public-private partnerships (PPP) and local participations. Of course, depending on the interests of the organization, the promotion of the technical focus started immediately. As usual, the discussions were heated.

Moreover, there was neither a pronounced environmental awareness nor a movement towards energy transition in the MENA region in 2009. After all, the visionary Moroccan government had already recognised the need to invest in renewable energies during the oil price increases in 2008 and developed ambitious expansion plans.

2.1.2 *The task*

During this time, the shareholders defined concrete tasks for Dii (Desertec Industrial Initiative 2009). A suitable regulatory and legal environment would be created for the exchange of electricity between the South and the North in the period from 2009 to 2012. In addition, a Rollout Plan was developed that served as a guidepost for investments, and considerations were made on reference projects. Dii also planned detailed studies on the feasibility of the Desertec vision.

The project was attributed great opportunities in terms of foreign and security policy: North Africa, which is regarded as unstable, would be tied to Europe in order to pave the way for greater stability and sustainability. The German political scientist and sociologist Claus Leggewie has already dealt with the expansion of the EU around the entire Mediterranean Sea. In 2009, the German magazine *Der Spiegel* even used the phrase anti-terror programme in an article in connection with the gigantic energy project (Pflüger 01.07.2009). He described the lack of prospects and the feeling of inferiority of many people in North Africa as the ideal breeding ground for terrorist recruitment. Words that almost ten years later sound more topical than ever in view of the refugee problem.

The Desertec vision countered this negative scenario with independence from fossil and nuclear resources. Conflicts over fossil resources and the spread of fissile material could be contained by their implementation. For the first time, the industry was also interested in a sustainable, carbon dioxide-free energy supply on an unprecedented scale.

In the case of the non-EU Mediterranean countries, which were almost exclusively dependent on fossil fuels, a high development and catch-up demand for energy supply was identified. The establishment of the private Desertec industrial consortium gave rise to the hope that renewable energy projects would become a regular business model and not remain typical development projects as in the past. Until now, they had been dependent on financing by development banks or official development cooperation.

Because mainly German companies had initiated Dii, German politicians had a great interest to help Dii on its way. Therefore, the German government promised all necessary support. The cooperation with the Federal Foreign Office, the Federal Ministries of Economics, Environment and International Cooperation as well as their institutions: the Chambers of Commerce Abroad (AHK), the German Society for International Cooperation (GIZ) and the Reconstruction Loan Corporation (KfW), developed quickly. On the EU side, there was committed support from Eurocommissioner for Energy and Transport, Guenther Oettinger. The support of other EU countries was still very moderate, a wait and see attitude. In MENA, the Moroccan and Tunisian governments were very interested from the beginning. The Arab League had expressed its interest from the beginning.

2.1.3 *The search*

In order to implement all these plans and intentions, a suitable manager had to be found for Dii. He should be able both to keep the large circle of shareholders with their different interests together, and to work with the governments involved in Europe, North Africa and the Middle East, generate income from partnerships, negotiate with civil society and lead a team of talented, highly motivated young people. This required a high level of technical, economic and social competence. The task turned out to be correspondingly difficult. Some of the shareholders were large companies in the DAX format of the German stock exchange with a wide variety of business objects, which made it difficult to act purposefully and in unison. The Dutch energy manager Paul van Son was chosen in September.

Figure 8: Paul van Son. Source: Dii

2.1.4 *The manager*

On October 31st, Paul van Son took over the position of managing director of Dii GmbH. His most important advocates were the technical director of ABB, Joachim Schneider, and RWE manager Fritz Vahrenholt, but also Gerhard Knies from the Desertec Foundation. Contacts that would later prove to be resilient.

The then 56-year-old manager had previously been head of Germany for the Dutch energy supplier Essent, which he had turned into the market leader in the wind energy sector. Further steps in his career were Siemens, TenneT, DNV/KEMA and Econcern. At Essent in particular, Van Son acquired a high level of expertise in electricity grids, energy trading and the sale and generation of renewable energies. In Holland he was called Mister Green Power, since he had acquired 800,000 customers for green electricity from 2001 to 2003. The *Financial Times Deutschland* introduced him in a frontpage article in October 2009: "Pioneers have no doubt that Van Son can pacify the diverging interests." (Gassmann 30.10.2011). "As a typical Dutchman, he is characterised by the polder model," described Jörg Spicker, Germany CEO of the Swiss energy supplier Alpiq, who worked with Van Son for many years as a German representative (Gassmann 30.10.2011). The polder model represents the basic attitude of many Dutch people to approach projects preferably in consensus with all the groups involved, but without ignoring any differences of interest. Van Son had also been involved in SEP (TenneT)'s 1989 plans to connect Holland and Norway via a power line in the North Sea.

2.1.5 *The team*

In December 2009, Paul van Son met Aglaia Wieland on the recommendation of Egon Zehnder. The then 35-year-old gave up her position as a consultant at the Boston Consulting Group to join Dii. In February 2012, she became second managing director of Dii. Wieland brought the mathematician Florian Zickfeld, also from the Boston Consulting Group, to Dii, who was mainly responsible for developing models. The young man quickly acquired the nickname whiz kid. In July 2010, Cornelius Matthes joined the company in the position of Director of Business Alliances. Matthes worked in a leading position at the Deutsche Bank in Milano, where he was responsible for organizing the conferences and helping to build the international network. In spring 2011, Klaus Schmidtke followed as press spokesman. The communications

Figure 9: The team of Dii. Source: Dii

specialist was sent to Dii by Munich Re, where he had already dealt with the MENA region. Philipp Godron was responsible for the electricity grids and had gained experience with Jordan at GIZ. Andreas Wischnat was part of the team as an analyst. He had studied in Lebanon and was increasingly involved for Dii in cooperation with strategic partners in North Africa.

Shareholders and partners of Dii, such as E.ON, Schott, Abengoa, Siemens and RWE, sent temporary employees to support Dii. Experts with roots in the region, such as Karim Asali (First Solar), Samyr Mezzour (Bearing Point), Ahmed Youssef (GIS Expert) and Maher Soyah (STEG) worked in critical positions in the team.

In the years 2010 and 2011, the Group was very dynamic and in a mood of optimism, which was due to the special task. Most of the employees were aware of the great cause they were involved in and the extraordinary support they received from business and politics. The enormous media attention was reflected in the team. In those days one or the other had the impression of being able to walk on air.

2.1.6 *The way is the goal*

Right from the start, Dii divided its tasks into three areas. Firstly, long-term strategic planning, the concrete implementation of the Desertec vision. In 2009, the Desertec Foundation presented to the public a fixed image of desert power, designed in DLR's 2005 studies. Meanwhile, things were already developing autonomously. More and more wind farms and hybrid gas-solar plants were planned and built in North Africa. The core task of Dii was de facto to arouse the public's interest in the long-term perspective, but it still didn't succeed. It continued to be regarded as the builder of large power plants in North Africa, which were inttered to secure Europe's electricity supply.

In addition, it was about the analysis and development of legal regulations and required start-up subsidies. The topic of cross-border feed-in tariffs vs. negotiable certificates for new plants was discussed intensively.

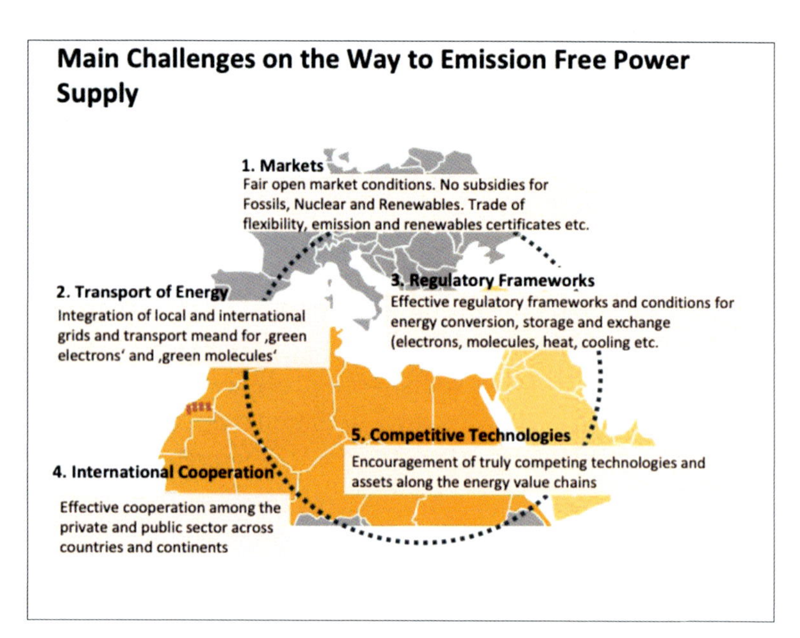

Figure 10: The tasks of Dii. Source: Dii

The third area dealt with proposals for appropriate reference projects planned in Morocco (1,000 megawatts), Algeria and Tunisia (500 megawatts each). In the first year of Dii, it was still open whether and how it would be involved in these reference power plants. The technology companies were predominantly in favour of a strong project development role, the majority of the shareholders saw Dii more as a trailblazer. In its own opinion, Dii, in cooperation with local and international partners, would explore the possibility of concrete power plants in Maghreb with an export option to Europe. At the end of 2010, the shareholders prevail: Dii would only act as an initiator and companion for reference power plants in the market. Companies from the Dii circle as well as external market participants were considered as builders of these projects. However, this decision could not correct the image of Dii as a project developer in the public view.

The financing of the 'Project' also attracted great attention time and again. In its initial studies, DLR assumed an investment sum of up to 400 billion euros for the Desertec project outlined there. Over an investment period of 40 years, this meant around 10 billion euros per year. In return, part of the region's energy supply and 15 percent of Europe's renewable electricity needs would be covered. After all, 350 million people lived in the MENA region at the time in an area two and a half times the size of Europe as a whole. Nevertheless, the figure of 400 billion euros was given an almost magical significance in the media. By way of comparison, a few years later Peter Altmaier, in his function as Federal Environment Minister, was to put the cost of the energy system transformation in the relatively small country of Germany at 1,000 billion euros. Anyone who is familiar with the energy supply of countries at the national level will confirm that no matter whether a traditional or modern, emission-free energy supply, the dimension of 400 billion euros over a period of 40 years in which just up to three investment cycles take place, is nothing unusual for MENA conditions.

2.1.7 *The dilemma*

Against this background, there was also a considerable potential for conflict between the shareholders with regard to the use of technologies. Especially the companies in the vicinity of the CSP power plants already built, such as Abengoa Solar, MAN Solar Millenium and Schott Solar, but also Siemens, which had just bought the CSP company Solel in Israel, had an understandable economic interest in building CSP power plants in North Africa as quickly as possible. However, solar thermal and photovoltaic systems were still very expensive in 2009. In the south of Spain, CSP was subsidised with about 27 eurocents at the time. In 2008 and 2009, Germany had planned similar feed-in tariffs for solar energy and significantly lower tariffs for wind energy. Therefore, there were considerations to proceed on the other side of the Mediterranean according to a similar principle – which turned out to be difficult to realize. Investments worth billions in initially still very expensive technologies in the region were considered rather questionable, although the World Bank, among others, already had a financing programme ready and waiting. Instead, a group of shareholders led by Munich Re advocated an approach that was open to technology and results. Detailed studies should first determine the appropriate technique. In addition, there were tensions due to the great cultural differences between the European and Arab countries. Dii promoted understanding and wanted to contribute valuable new insights with reference projects. During this time, Paul van Son saw Dii as having a duty to create good conditions for the region's own long-term development of renewables. The region would see Desertec as one of its children.

Additional conflict material was provided by the new EU guidelines for renewable energies, which had emerged in the wake of the Kyoto Protocol and the legally binding obligation to reduce greenhouse gases. The guidelines allow the promotion of electricity exchanges from renewable sources across the borders of certain EU countries. The regulations known as Articles 6 and 9[1] seemed rather market-agnostic. A simpler, market-oriented solution, the use of negotiable Renewable Energy Certificates (RECs) in the EU Member States and their non-EU neighbours, had already been proposed by the EU Commission in 2008, but had been rejected by Germany in particular. Instead, Germany relied on the feed-in tariff it had developed.

---

1 Article 6 of the Directive provides for a regime to enable the exchange of electricity from renewable energy sources. It gives member states the possibility to regulate the transfer of renewable sources. Article 9 includes third countries. Thus, the Directive could, in principle, be the basis of legislation regulating the transport of renewable energy from the desert countries to Europe. The regulations were impractical and legally questionable. Article 6 has hardly been used in Europe until now (2018), Article 9 has never been used.

Expert voice: Ernst Rauch (MunichRe)

Ernst Rauch. Source: Munich Re

Ernst Rauch, Chief Climate and Geo Scientist, is responsible for Climate & Public Sector Business Development worldwide at Munich Re.

Covering risks from natural catastrophes has an exceptionally long history for Munich Re as a reinsurer. The story begins in 1906, when Munich Re assumed substantial claims payments for the severe earthquake in San Francisco. In the 1970s, an expert unit of geoscientists was set up and a worldwide database with detailed natural catastrophe loss information was developed. In the following years, the data collected revealed noticeable changes in loss frequencies that could hardly be explained without a contribution from climate change and the associated extreme weather conditions.

The reinsurer was increasingly concerned with the question of how it could use its data to inform about this development and, on the basis of its claims expertise, help reduce future catastrophe losses. Without measures to improve natural catastrophe resilience, it can be assumed that climate change will lead to an increase in personal injury and property damage in the medium and long term.

One of the responsible persons at the Munich company from the financial sector is Ernst Rauch. The geophysicist has been working on climate change at Munich Re for around 30 years. He has also accompanied Dii since its inception.

Commenting on Munich Re's commitment, Rauch said: "With our risk management competence, we support technologies that lead from carbon-based technologies to decarbonised economies." This is called 'Greentec' in New German and is closely linked to renewable energies. The reinsurer is thus networking the topics of climate change and new technologies/renewable energies. The company from the financial sector established ties with the Desertec Foundation in 2008 through talks with Gerhard Knies.

An examination of the DLR solar energy studies at the time strengthened the conviction that the vision of desert power for the MENA region and Europe could be part of the solutions for avoiding a rapidly advancing climate change. This was in line with the reinsurer's findings.

In 2009, the DAX-listed company from the financial sector was not only interested in climate and energy, but also in social and societal perspectives. Rauch finds it regrettable that the Desertec University concept did not really come to fruition. This should educate people from both continents for the purpose of knowledge transfer about renewable energies in the direction of the MENA region, also an idea of Gerhard Knies.

During this time, the Desertec theme is moving up the agenda at Munich Re's meetings, and Torsten Jeworrek, member of the Board of Management, is getting involved in the founding phase of Dii. In the late spring of 2009, contacts with other industry representatives were established with the weight of his vote on the Executive Board.

First of all, a basic concept should be developed. Some of the initial 12 companies delegated representatives to Dii for this purpose. "This was exactly the right way for us," says Rauch, commenting on the plan, "because the first step was to clarify the scientific and technological issues of electricity generation and grids, as well as possible financing systems. However, the public was given the wrong impression that Dii was sitting on 400 billion euros and would act like a construction company."

Rauch also complains about the recurring focus on technology within the framework of the procedure. DLR's preparatory work is closely linked to solar thermal energy, which corresponds to the state of the art in 2005. This then settled in the minds. It was true, however, that Dii always had the goal of being technologically open.

Different energy technologies generate electricity at different locations at different costs and load profiles. At the daily level, solar thermal energy and PV are better at the load curve than wind energy, because they produce electricity when demand is greatest. Rauch sees the restriction to individual technologies as a weak point, similar to the restriction to 400 billion euros. The work on solar thermal energy originated in the DLR studies around 2005 and was decisive for the initial developments of the Desertec concept. The investigation of further technology questions, as well as the treatment of market questions was then the task of Dii, according to the representative of the reinsurer.

Munich Re was generally positive about a reference power plant in the MENA region. It could have demonstrated that such projects are also feasible under local conditions. In the sense of a market model for electricity exports to Europe, however, a reference power plant did not function at that time because the supply of electricity from North Africa to Europe did not fit in with the existing market conditions. For example, Morocco was an electricity importer from Spain because its own generation capacities were not sufficient to cover its nationwide demand. There were different positions in Dii on the import/export issue. Rauch commented: "A reference power plant with substantial support from KfW Bank would have initially supplied energy for the local market." A power plant with electricity exchange in accordance with Article 9 of the EU Directive for Renewable Energies would not yet be possible in today's market structure. In the meantime, several solar power plants (CSP, PV) have been developed by the Moroccan solar agency MASEN, which at least in some areas came close to the Dii concept for reference projects to implement the Desertec vision. Dii shareholder ACWA-Power was significantly involved in the realization of these projects. With regard to the conflicts sparked off by the discussion about the reference power plant, Rauch believes that such very ambitious projects also involve people with different personality structures. "This can also complicate the implementation, but in the end, it is part of the reality of life. For Munich Re, the studies carried out by the Dii team together with external research institutes are a great success. They answer technical questions regarding power generation from renewable energies, power grid structures and market conditions in the target areas."

In 2018, Rauch sees the situation around desert power changing, as developments in the area of migration have their influence. Desertec offers development opportunities for North Africa. This brings politics back into the Desertec vision of linking Africa and Europe more closely through joint renewable energy projects, so that an economic perspective is created for the large, young population. Increasingly, it is no longer just a question of electricity exchange between Africa and Europe, but also of a possible production of synthetic fuels such as hydrogen and methane as well as methanol through renewable energies. The basic concept would change, not electrons but chemical energy carriers could now be transported. If Germany is to be 95 percent decarbonized, the import of synthetic energy sources is needed. "That's where MENA is a potential target region," says Rauch.

2.1.8 *New solutions*

This interdisciplinary and intercultural task was to be solved by the team around Paul van Son and a handful of external consultants. The mostly young Dii team members often brought with them extensive experience from management consultancies and large companies. However, they first had to familiarize themselves with the complex issues of the energy industry and the various technical possibilities. The group worked under the watchful eye of the public and the corporate bosses on a large number of open questions.

The first example is the electricity grids. Technically speaking, it is possible to transport electricity over long distances, i.e. several thousand kilometres, safely and economically – over land, in the ground or under water. The HVDC is used for long distances and large volumes. In principle, it can be well connected to the existing alternating current grids in Europe. Modern examples can be found in China and other countries that already have DC lines with a length of up to 2,000 kilometres in operation. Electricity exchange between continents has also already been implemented. For many years there has been a 600-900 megawatt AC connection between Morocco and Spain and an up to 4,000 megawatt approved connection between Bulgaria/Greece and Turkey. Although international electricity transport is not a particularly technical problem, new lines, especially in densely populated areas, are difficult for residents to accept. In some countries, including Germany, new transmission lines are hardly justifiable. Instead, they switch to more expensive underground cables. In Germany it was already assumed at that time that nuclear energy, lignite and hard coal would no longer be permitted at some point and would have to be replaced by renewables. However, a decentralised energy supply would only partially compensate for the deficit and, in the view at the time of energy supply that was still very much based on central generation, too little or no base load generation, Germany would be dependent in the long term on electricity imports from abroad and the electricity should be 'green', it was said. However, there is only one type of physical electricity that can be exchanged over the interconnected networks. Green electricity is therefore not a physically negotiable product, but the composition of physical electricity combined with proof that the electricity was generated somewhere by a renewable plant. A fact that always leads to too much discussion and, above all, misunderstandings.

How to get electrical power from the deserts to Europe?

A possible solution was the export of electricity, attributed to solar thermal power plants in North Africa, to Europe, in particular to Germany. The idea was supported by the solar thermal industry as it would create a market for (German) solar thermal technology in North Africa. But the financing of such power plants was problematic. Where would the funds for the power plants in North Africa come from? During this time, the idea was born to qualify electricity from solar thermal plants for similar feed-in tariffs to those applicable in Germany.

From a political point of view, however, this was only conceivable if the electricity from such plants actually flowed physically to Europe, or rather comprehensibly to Germany. Dii discussed two possible solutions: DLR proposed that the power plant to be promoted would be connected directly to the European grid or even directly to the German grid by its own pipeline. However, this approach appeared to be prejudicial in all respects and impracticable. It was more feasible to feed the power plant to be promoted into the local North African grid and to ensure that it was connected to Europe. In theory, the current could then find its way to Europe. In the event of bottlenecks in the local network, exports would have to be reduced or interrupted in order to ensure local supply – not simply implemented. A second problem was that it was not possible to distinguish between renewable and conventionally generated electricity when it was fed into the European grid. Under these circumstances, how can it be ensured that the new green power plant is specifically supported and not a coal or gas-fired power plant?

In principle, two different ways of promoting renewable energies in North Africa were considered possible: either investment grants, reduced-interest loans, donations and similar measures for the construction of renewable plants. This is linked to the obligation to export electricity directly or indirectly to Europe or Germany. Or a feed-in tariff for every kilowatt-hour of electricity that ends up in Europe from the renewable power plant and is taken from the local grid.

The conceivable case of simultaneous surpluses in the European grid and bottlenecks in the North African local grid was more the case than the other way round. In this scenario, electricity does not flow from the power plant to be promoted to Europe, but from Europe to North Africa. Even if the state-run energy supply systems in North Africa did not yet have an open pricing system, the system operators would make full use of the opportunities for cheap imports. The Moroccan system distributor ONEE, for example, traded on the Spanish electricity exchange OMEL.

There were other reasons against a rigid link between subsidies and transport to Europe. For countries such as Egypt, which did not yet have a network connection with Europe, physical feed-in to Europe could not have been promoted in the same way. In addition, wind and photovoltaic (PV) systems would always be at a disadvantage compared to solar thermal systems because their power generation is not controllable (dispatchable). Even the integration into the local network poses special challenges. The transmission of electricity from wind or PV plants to Europe would require very complicated regulations.

But there was a more elegant solution: negotiable green certificates (e.g. I-RECs, International Renewable Energy Certificates). A generator, green or not, connected to the grid delivers physical electricity like any other. In Europe, electricity from plants that produce CO_2 emissions is punished by the ETS (Emission Trading System). However, this does not apply in the MENA countries. The electricity from green areas is net emission-free and this would be rewarded if possible. For example, by allocating a green certificate on the basis of a formally established Guarantee of Origin (GoO) for each kilowatt-hour of green electricity generated. All in all, a green power plant generates quite normal electricity plus a green certificate per kilowatt-hour. This is a great advantage in the international electricity market, as electricity can be negotiated across borders as normal. In addition, the green certificates are traded on a separate market. With the help of green certificates, the end consumer can obtain green electricity everywhere, composed of electricity from the grid and a certificate. With this model, there would be no spatial limits to the promotion of renewable plants.

For such a model to work, all participants must agree on a standard (e.g. I-RECs) and seamless monitoring. This requires watertight international rules for registration, supervision and trade. Today blockchain technology offers a good solution. Green certificates could be negotiated and marketed as green coins within this framework. Governments could buy I-RECs for promotion and/or set minimum requirements for green electricity from customers, thereby giving credible value to I-RECs.

2.1.9 *First beginnings in Maghreb*

Although all of MENA was the object of study and planning, the focus in the early years was clearly on the Maghreb countries (Morocco, Algeria and Tunisia). When Dii first sought talks with those responsible there, they were welcomed with open arms, but also had to dispel reservations.

Morocco

A delegation of Moroccan Energy Minister Amina Benkhadra, Abderrahim El Hafidi, Director for Electricity and Renewable Energies at the Ministry of Energy, Mines and Sustainable Development, Moustapha Bakkoury, head of the newly founded MASEN, an organization for the implementation of solar energy in Morocco, and Ali Fassi Fihri, head of the Moroccan Ministry of Energy, met Van Son and representatives of the German government in Berlin in December 2010. The atmosphere was euphoric and the expectations of a cooperation were high.

The ambitious King Mohammed VI planned early on the implementation of renewable energies in Morocco. The most western state in the Maghreb, with its strategically important position on the Atlantic and the Mediterranean, had to import expensive oil, gas and even hard coal from South Africa to generate electricity. He has also been importing electricity from Spain for many years. The high , volatile costs were regarded as an obstacle to the planned development of the Maghreb state. Therefore, the country, which already had solar and wind plans for 2,000 megawatts ready and waiting, sought cooperation with Dii. But it wasn't clear yet whether the Moroccan plans and Dii's ideas for a reference power plant would fit together. MASEN, ONEE, the Ministry of Energy and other authorities were sceptical, but also curious what the Germans were up to. From the beginning, they worked friendly with Dii, supported by the German government. All participants quickly agreed on a formal cooperation, including the Moroccan King, Bakkoury and Van Son.

Algeria

The Algerian government initially showed no interest in cooperating with Dii. The Minister of Energy, Chakib Khelil, made it clear through the media that foreign groups would not try to build power plants in Algeria to meet their energy needs. He was angry that Algeria had not been invited to talks in advance. In his view, such an initiative should be politically endorsed by both sides. But he was willing to talk as long as Dii promised to support technology transfer in partnership with Algerian companies. The Algerian family business Cevital was already a shareholder of Dii and ready to do so.

But there was no more exchange with Khalil, because shortly afterwards Youssef Yousfi took over his ministerial position. In spring 2010, Töpfer and Van Son held the first talks with the Director General for Energy. It would be a while before the ice broke. First of all, the Algerians complained that Dii companies had already inquired about the possibilities for CSP technology in Algeria. The choice of the energy mix is not the responsibility of Dii, but of the Algerian government and industry. At the end of 2010 President Abdelaziz Bouteflika paid German Chancellor Angela Merkel an official visit to Berlin. The topic of Desertec was mentioned positively in the margin. Shortly afterwards, Dii was invited for an interview with Minister Yousfi. The go-ahead was given for close cooperation with the Ministry and with Sonelgas, Algeria's state-owned electricity and gas supplier. It is remarkable that the population had high hopes for Desertec. They expected a *deus ex machina*, the early construction of huge power plants for export to Europe and new jobs for young people. The idea is still anchored in many Algerian minds in 2018, but with a bland aftertaste. However, the colonial legacy of Algeria and France is a burden on relations between North Africans and Europe. The memories of the injuries of the Algerian war are still alive. The military enjoys a high status and the key positions of government, administration and army are occupied by veterans from the Algerian war. At the beginning of 2019, there appears to be a change here.

Tunisia

The Tunisian government was initially reluctant to cooperate with Dii. Already in December 2009, Van Son met with Energy Minister Chebli, State Secretary Rassaa and representatives of the state-owned energy supplier STEG – Tunisia was already discussing plans for renewable energies at this time. In 2010, an agreement was reached with the Tunisian energy supplier STEG to cooperate in the system studies.

2.1.10 *Connecting*

In the meantime, Van Son began to build an international network of representatives from politics and the public and visited the governments of many participating states as well as the EU Energy Commissioner Günther Oettinger. He wanted to support a reference power plant and to transport the resulting electricity from Morocco to Europe. Important players, influential people in Europe, North Africa and the Middle East had to be inspired for the still German-dominated initiative. The impressive and marketing-effective images of the Desertec vision were to be brought closer to the people. In his presentations from this time Van Son emphasizes an important context: Connecting Energy = Connecting Continents = Connecting People. The key to this connection was intercultural communication.

2.1.11 *The first annual conference*

The first annual Dii conference was held in Barcelona in autumn 2010. It brought together important representatives and interest groups from Europe, North Africa and the Middle East. Energy Commissioner Oettinger was also present. At the conference, Dii tried to make it clear that it is not a project developer. It presented the work programme and the ongoing system studies that it carried out with the Fraunhofer Institute. The first task was to arouse enthusiasm for the desert power project in North Africa and the Middle East. From the southern shores of the Mediterranean there were, as already described, different, but mostly reserved, sometimes negative signals and again and again the accusation of neo-colonialism arose. With the Desertec idea, another European player intervened in the developments of the Maghreb from a North African perspective. Even if the signs were different this time, there were still many reservations. Moroccans and Algerians fear that they will become dependent on European technology and the supply of spare parts; colonial conditions in a 21st century garb adapted to green needs. Subsidy funds from Europe were welcome, but Europe would not give the impression of wanting to dominate. A difficult task for Van Son and his team right from the start. It was important to build bridges instead of preaching German ideas.

Expert voice: Professor Jochen Kreusel (ABB)

Prof. Jochen Kreusel. Source: ABB

Jochen Kreusel, Market Innovation Manager of the Power Grids Business of the ABB Group, represented the provider of power grid technologies at Dii.

In my opinion, Dii came about rather surprisingly. The famous headline of the *Süddeutsche Zeitung* forced the negotiating partners to act quickly and generated a lot of public attention.

Desertec was committed to neutrality with regard to production technologies from the outset, even though it was initially frequently perceived by the public as a solar thermal project. On the one hand, this was due to DLR studies, which were the origin of the Desertec idea, and on the other hand to the fact that manufacturers of solar thermal power plants were initially more strongly represented in the consortium than those of the other generation technologies.

ABB was one of the first Desertec shareholders, had early contacts with many players in the network that ultimately led to the Desertec Foundation, and provided technical support. Our specialists for power grids therefore enjoyed working with the established organisations and the foundation.

An important motivation was to expand the market for our solutions in accordance with our corporate philosophy – *Power and Productivity for a better World*. So, it wasn't just about future sales, it was also about positioning our brand.

I consider the fierce discussion about the investment sum estimated at 400 billion euros to be a blunder in the start-up phase. The number was not based on reliable data, nor did the consortium itself plan to invest. The fact that no clear position was taken at this point gave rise to false public expectations, which were later disappointed.

The first meetings of the partners are much more positive in my memory. Although they came from completely different industries and largely did not know each other, a constructive working atmosphere developed very quickly, so that we were able to agree on a well-formulated strategy after only a few weeks: Based on the potential of a joint use of renewable energies in North Africa, the Near and Middle East and Europe, suitable framework conditions and a market for investments in renewable energies and the associated infrastructure would be created.

In addition to the extensive studies and country analyses of the first, three-year work phase, the possibility of a reference project was also part of the strategy. Such a project would have made it possible to study the challenges and difficulties involved in electricity transmission to Germany, and the pitfalls for politics would also have become visible. The topic of electricity transmission through Spain and France was already quite far advanced, but the project did not (yet) fit into time: Spain already had a surplus of renewable energy and insufficient transmission capacity to France.

The importance of the reference project and its failure was clearly a critical point in the history of the consortium. Although such a project was only an option from the outset, it was perceived by the public as the essential objective. In addition, manufacturers of solar thermal systems were under increasing pressure to demonstrate the efficiency of their technology in view of the dramatic drop in the price of photovoltaics. Ultimately, this led to tensions that were also registered in public. However, it can also be seen as the merit of the consortium that it survived this crisis and even completed a second, two-year work phase, which was not originally planned.

In the beginning, the Dii Group's knowledge of energy economics was not particularly great, with the exception of the energy suppliers, such as RWE and E.ON. The DLR study from 2005 also does not yet deal much with energy industry issues but is limited to macroeconomic considerations. In retrospect, that was a shortcoming.

Today, at the beginning of 2018, I consider the project to be interesting for two reasons: When comparing the best locations for renewable energies in Africa and Europe, there are actually no major differences – at least not so great that they would justify the transport effort over thousands of kilometres. What is different is the availability of quantities:

Europe is densely populated with a limited number of such good locations. However, this effect only becomes apparent when the whole of Europe has reached a share of renewable energies beyond the 20-20-20 targets, i.e. beyond 30 percent of renewable energies in the electricity sector. For the current energy policy objectives, this is the case no later than 2030. In many places this scarcity situation is already visible today. In pioneer countries for wind energy, such as Great Britain, Denmark or Germany, new areas for wind energy are now almost as unpopular as overhead lines.

The second effect for the energy sector is the combination of seasonal complementary supply and load profiles – an effect of the wide-ranging interconnection that both regions need. In the European debate, the control of seasonal fluctuations is increasingly emerging as the central problem of energy system transformation. A geographical expansion of the network can make an important contribution to this, as the Dii studies have clearly shown. Desertec is therefore an affordable option that is coherent, technically viable and would contribute to the supply of electricity to the regions.

Even though an electrical connection between Africa and Europe will only be needed in a later phase, in the near future renewable energies will be needed in North Africa and the Middle East primarily to cover the growth in consumption there. However, planning for the intercontinental interconnection would start as early as possible and create the necessary conditions, because the development of a cooperative culture in energy supply takes a lot of time. The project is once again high on the political agenda.

A standard argument against Desertec is that energy supply depends on unstable regions. However, when it

comes to fossil energy supplies, we only rely on supplies from a few countries and regions, and our dependence on unstable regions is also high. In the case of gas, there are also few pipelines between producers and consumers.
On the contrary, the Desertec concept proposes hundreds of connections. Desertec is also an opportunity to stabilise economic cooperation on the southern border of the European Union – ultimately this is the only way to tackle the stability problem. It has been proven that demarcating oneself does not work – just think of the current refugee crisis.

The urgently needed process of growing together between Europe, North Africa and the Middle East is long and difficult. Here it is worth taking a look at Europe, which today is generally regarded as stable: this stability followed centuries of constant warlike conflicts in our country as well.

The tense relationship between the Industry Initiative and the Desertec Foundation was a missed opportunity in both directions. On the positive side, it must first be noted that the rather industry-sceptical Foundation and the large industry have come together in early phase. However, most of the shareholders and the foundation did not see the opportunities that this offered; they felt that the cooperation was more like a liaison in which agreement had to be reached somehow. But that wasn't enough to stay together in the long run. On the one hand, the Foundation, which exclusively prioritised the fight against climate change and claimed communication sovereignty over the entire Desertec issue; on the other hand, the companies, which have to survive in a multidimensional competition on a daily basis. Although both sides, and Dii itself in particular, have made efforts to keep the Foundation and the industry together, they have ended up going their separate ways for lack of a common vision of collaboration.

Overall, the story of Dii is still a success story. Our studies are unique. In 2009, only Morocco had plans to implement renewable energies. In the meantime, all the countries have issued something on the subject. This development is not only due to Dii, but the industry consortium has catalysed this process and given credibility to the issue. In addition, the composition of the shareholders has become more international. In the beginning there were 12 German companies and the Desertec Foundation, but soon there were 20 partners. More than half of them came from an international environment. A very positive result. ABB has learnt a lot during these years and continues to work on these topics – partly with the partners of the still existing Dii.

2.2 *The year 2011: The search for suitable answers*

Van Son had realized early that Desertec should be the child of local governments and industries: The North Africans wanted to go their own way and not be dependent on European know-how. It was not primarily about electricity from the desert for Europe, but about electricity for the region. Export opportunities would only be exploited when energy prices in Europe would at some point rise well above the production costs in Morocco. Dii would only pave the way for all stakeholders. Projects planned by local developers would therefore be supported as far as possible and would be a reference or learning object for the region.

2.2.1 References for the market

A first reference power plant with a capacity of 580 megawatts was planned in Ouarzazarte, an up-and-coming provincial capital about 200 kilometres from Marrakech. A prestigious project consisting of four solar power plants with different technologies, which were to be compared and evaluated in practice. International development banks were able to raise the necessary funds.

The city is located on a plateau about 1,160 meters high, very convenient from a traffic point of view, and has about 150,000 inhabitants. It had grown rapidly in previous years. Approximately 12 kilometres before its gates, surrounded by small villages, there was an area suitable for solar thermal power plants, according to expert opinion. The Saharan continental climate of the interior guaranteed solar radiation with an energy of 2,600 kilowatt-hours per square meter per year. It is about 2.5 times as much as in Amsterdam. A nearby dam was able to supply the water for cooling purposes, for efficient operation of the solar power plants. Unwanted disturbances by flora and fauna were hardly to be feared on the barren land.

Ouarzazarte is a modern Muslim town with vibrant life around a central square with a well-developed infrastructure. Due to its special lighting conditions, it accommodates film studios for monumental films with an oriental backdrop. However, the location is not quite perfect: it lies in the southern central part of the country, far away from centres such as Tangier, Rabat and Casablanca. The way there leads over winding roads through the Atlas Mountains, not optimal for the transport of bulky, technical equipment. Should difficulties arise, experts would have to be flown in from Casablanca at great expense.

Two solar thermal parabolic trough power plants and later also a tower power plant as well as PV systems were planned. Construction would start at the beginning of 2013. The world watched the project curiously. It also served as an initial inventory of Moroccan industry; the aim was to gain an overview of the various players. From the outset, Morocco had an interest in carrying out as many value-added processes as possible in its own country, in order not to replace the existing dependence on imports of fossil raw materials with a dependence on European technology. This also included a more detailed analysis of the various value-adding steps of solar technologies, which was to be prepared as presentation material. The Moroccan Agency for Solar Energy (MASEN), founded by law in March 2010, was responsible for implementing the project. A commitment that fell on fertile ground. The Moroccans liked to be pioneers of a worldwide energy revolution and to occupy a leading position on the African continent alongside South Africa.

Mustapha Bakkoury was responsible for Ouarzazarte, a confidant of King Mohammed VI, who had worked on the 2-billion-euro container port project in Tangier. There he distinguished himself as an excellent project developer and gained experience in financing major projects through development banks. The dynamic man was politically well integrated and possessed the confidence of the King. He was also regarded internationally as the ideal candidate for this position.

Figure 11: Mustapha Bakkoury and Paul van Son. Source: Thomas Isenburg

Bakkoury and his team succeeded in convincing the counterparts in the region that they were not primarily interested in German interests, but in progress for the region, through many conversations at different levels and with the help of information brochures in English, French and Arabic. It also motivated the young Dii team to change their perspective.

However, the discussion about reference projects and the fact that Dii itself did not act as developer or investor led to internal tensions. At the same time, the Dii team is becoming more international. Managers and experts from the Arab world, such as Karim Asali, Samyr Mezzour and Achmad Youssef, follow. With them a lot of intercultural competence came into the team. At the same time some employees of Dii tried to learn Arabic and French.

Dii was also supported by the German Foreign Minister Guido Westerwelle. Van Son accompanies him to a Bertelsmann Stiftung conference in Rabat in May 2011. In his contribution to the conference, Westerwelle particularly emphasized the opportunities of a solar energy partnership and invited Dii to hold its conference at the Federal Foreign Office in 2012.

Communication remained a central task of Dii. Again and again, interviews with Van Son and reports about Dii appeared in the big newspapers. The team travelled the region together and also provided coverage in Arab media, including a four-page interview with press spokesman Schmidtke in an Algerian newspaper (Lahdiri 2011). There was euphoria in the team around the Dutch manager, more young employees joined in. Small groups around the key people Aglaia Wieland, Cornelius Matthes, Florian Zickfeld, Philipp Godron and Klaus Schmidtke, worked daily until late in the evening with a lot of commitment. Various discussion rounds with the bosses of DAX companies, the former BND boss Ernst Uhrlau and other industry leaders signalled the importance of Dii.

2.2.2 *The difficult relationship with the Desertec Foundation*

The relationship between Dii and the Hamburg-based Desertec Foundation was complicated. The objectives of civil NGOs and industry representatives differed considerably. The Desertec Foundation felt itself to be the guardian of the idea of Knies and the results of the DLR studies. In contrast to the shareholders and partners of Dii, it was unfamiliar with market economy ideas, which among other things led to different communication strategies.

Despite similar objectives, Dii and the Foundation did not match well. Even though both saw themselves as non-profit organizations, they increasingly kept their distance. This had already become apparent when Dii was named, because the Desertec Foundation did not make the brand name *Desertec* available, which had quickly gained wide recognition through the participation of numerous companies in Dii. Dii was almost relieved about this, as it was able to develop in a market-oriented manner independently of the Foundation and rethink initial assumptions of the Desertec vision. However, the public continued to see Dii as a representative or implementer of the original Desertec idea, which would lead to even more misunderstandings and ill feelings.

The energy manager Van Son and the father of the Desertec idea, Gerhard Knies, had a good relationship and similar goals; but Knies, who himself was only marginally connected to the Foundation, was also alien to market ideas. Van Son as a pragmatist, on the other hand, was primarily concerned with the realisation of sensible projects in the region. Despite many attempts, neither they nor other representatives of both organizations succeeded in persuading them to cooperate fruitfully. In 2013, the Foundation would withdraw as a shareholder of Dii.

2.2.3 The Arab Spring

The Arab Spring of 2011 played into Dii's cards, giving European investors hope. The demonstrators' demands for work, participation and social justice made access to the market more likely.

Figure 12: Especially young people demonstrate at the Arab Spring. Source: Thomas Isenburg

As a result of the protests, Mohammed VI recognized the Berber language Tamazight in Morocco and transformed the Moroccan government into a constitutional, parliamentary, democratic and social monarchy. Many of the demonstrators in Morocco were unemployed academics who now received public service jobs. However, the serious injustices remained. Subsidies from America and Europe continued to flow into the country, which is poor in raw materials.

Not in all countries did the protests remain as moderate and peaceful as in Morocco. They took on forms similar to civil war and countries like Syria and Libya sank into a chaos that still persists today. Refugee flows burdened the region. Lebanon, Jordan and Turkey were particularly affected, as were European countries to a lesser extent.

Van Son nevertheless saw the work of Dii unaffected by the social upheaval in North Africa and the restlessness that also emanated from it in Europe. Energy transition and the development of desert power do not depend directly on how states are organised, but on the needs of the regions and the technologies available. Progress is possible both bottom-up and top-down. In an interview with the *Neue Züricher Zeitung*, the energy manager welcomed the transformation process as a positive sign that the population in some countries now has the opportunity to organise themselves democratically. In the same interview, Van Son also bet a bottle of Barolo that Desertec's overall report would be available by 2012 (Im Winkelried 18.04.2011). However, the former Vice President and Executive Director of the World Bank as well as Secretary of State for Finance of the German Federal Government, Caio Koch-Weser, stated in the *Frankfurter Allgemeine Zeitung* that he expected developments to be delayed because decisions had been postponed (Herrmann 04.11.2011).

Even though the Arab world did not show any significant structural changes in the years after the Arab spring, it had landed on the radar of the industrial nations.

2.2.4 Fukushima and the energy system transformation in Germany

The expansion plans for renewable energies in Germany were already laid down in the German government's 2010 energy concept. With the reactor accident in Fukushima, the phase-out of nuclear energy came even more into focus. At the same time, the media lost interest in the export of electricity from the

MENA region to Europe. They are now rushing to the development of renewable energy. Germany faced a rapid development of renewable energies and said goodbye, albeit hesitantly, to fossil fuels. Now they wanted to generate renewable electricity themselves as far as possible. In addition, consumption stagnated. The gradual shutdown of the nuclear power plants did not prevent the existence of significant generation overcapacities. The other EU countries had a hard time shutting down their nuclear power plants. Each country chose its own path and few synergies were used.

From now on, Germany wanted to become a world laboratory for renewable energies. An ambitious project. Wind and PV plants were built in many places, but the expansion of the power grid lagged behind. Network optimisation opportunities in the European context were hardly sought. In the north, for example, large capacities for wind power plants were created, but their electricity could not be efficiently transported to the consumption centres in Europe and the centre and south of Germany without clever overall coordination (Franconia, 2013). There was also resistance to wind power plants amongst the population, especially in rural areas; they would disturb the landscape and generate noise.

At the end of 2011, the market for PV systems changed considerably. To date, manufacturers of photovoltaic modules have benefited from the generous EEG subsidies, especially in Germany. Now the sky over the German PV industry has clouded over considerably. Chinese suppliers bought the German expertise and built huge factories for the production of solar cells and modules. Due to low wage costs and energy prices and sufficient capital resources, they could produce cheaper than their German competitors and brought large quantities of PV systems onto the market at an enormous speed. This was followed by a series of takeovers and bankruptcies of German companies. The

CSP industry was surprised by the rapid drop in prices. At the end of the year, one of the Dii shareholders, Solar Millennium AG, was also hit. The company was one of the German pioneers in the field of solar thermal CSP power plants and had built the Andasol power plants in southern Spain. This technology was to be widely used within the Desertec framework.

2.2.5 *The second conference in Cairo*

Despite a mixed mood in Germany, the mood was good at the second Dii conference in Cairo in November 2011. A sense of optimism could also be felt on the Nile with a view to the Arab spring. Some oil companies expressed their interest in investing in renewable energies. The governments of the North African states were more interested than in the previous year. Dii was looking even more actively for a way to cooperate directly with North African governments and arouse local interest in renewable energies – actively supported by the German Federal Government.

The conference brought together 300 experts and decision-makers from around the world who had a much wider agenda than 2010 to work through. Technological progress and initial experience with plants in the desert provided enough impetus. Examples were the 100-megawatt CSP Shams power plant in the United Arab Emirates and the 500-megawatt Safarana wind farm in Egypt. With ministers, state secretaries and directors from Egypt, Germany, Jordan, Morocco and Tunisia, politics was well represented. The conference programme included a visit to a CSP hybrid power plant. These power plants produce electricity from solar radiation (during the day) and natural gas (at night). The atmosphere was peaceful and constructive, and the Desertec Foundation was also warmly involved. But as in the desert, the silence lasted only until the next sandstorm.

3. The years of decisions (2012-2013)

After a euphoric start, Dii had taken up the real work. With the first results of the largest infrastructure plans of the past decades, tensions between the players also increased and turbulent months followed.

3.1 *The Desertec Foundation and TuNur*

For Dii, the year 2012 began with a bang. A first Desertec project is to be developed in Tunisia, the *Frankfurter Allgemeine Zeitung* reported in January (Von Hiller 24.01.2012). The British consortium TuNur, around businessman Kevin Sara, planned the construction of a CSP power plant with a capacity of 2,000 megawatts. The announced capital requirement was up to 11.5 billion euros, of which some 1.5 to 2.5 billion euros was earmarked for a submarine cable between Tunisia and Italy. This was announced by the Desertec Foundation at an event organised by Frankfurt School Verlag. Dii was not involved in the project, although TuNur was only an associated partner of Dii – in retrospect, the break between the two institutions was already apparent. At the time, however, Thiemo Gropp, Managing Director of the Desertec Foundation, was still confident that the Foundation would continue to work with Dii.

The TuNur project was based on plans developed in cooperation with the Tunisian engineering and planning office Comete Engineering, and after the Arab spring, further concretised with Tunisian entrepreneurs, in particular Fethi Somrani from Top Oilfield and Cherif Ben Khelifa, the former manager of Lundin Tunisia. Dii did not consider TuNur's project to be particularly promising after having discussed it with its representatives. The experts doubted above all the dimension of the complex: 2,000 megawatts corresponded roughly to the output of two nuclear power plants in a country like Tunisia, which only had a peak load of around 3,000 megawatts. The necessary investment sum of around 11.5 billion euros was correspondingly offset by Tunisia's gross domestic product of around 40 billion euros. The planned site in the south-west of Tunisia at the edge of the Tunisian desert was hardly developed, but was characterized by an advantageous solar radiation, which was up to 30% higher than the Spanish sites. As the region lacked water, the technical planning was designed from the outset for a CSP tower power plant which, thanks to air cooling, manages without cooling water and thus requires 90% less water than a CSP parabolic trough power plant. The estimated electricity production costs of approx. 12 to 15 eurocents were at the level of European offshore wind turbines.

An exclusive cable connection to Italy was necessary because, among other things, those responsible were expecting only low revenues from feeding into the local network. The gas pipeline, which supplies Algerian gas to Italy via Tunisia, served as a model. Within the framework of this pipeline, Tunisia had secured the right to use part of the gas supplies for its own energy supply – a similar right was also part of TuNur's plans. The energy company hoped for subsidies or feed-in tariffs from Italy or another European country, where market prices were negotiated on average between 3 and 6 eurocents per kilowatt-hour. However, the cost of subsidised electricity from offshore wind farms in the North Sea at that time was still 15 eurocents per kilowatt-hour, so a feed-in tariff of this size was theoretically conceivable. In 2012 there was a general overcapacity in Europe and at certain times market prices were zero or even negative. It was not foreseeable in 2012 how prices and production costs would develop in the coming years. In the case of wind and solar energy, however, a further trend towards cost reductions was observed in Europe, well below the general market price, which was determined by fossil energy. This made it difficult to imagine that the Tunisian

CSP plant itself could be successful, even with subsidies. From today's perspective, in 2018, however, it is worth pursuing the project idea further, as the costs for CSP technology are now considerably lower. Depending on irradiation conditions, records of up to 7 dollarcents per kilowatt-hour are reported in Dubai. A CSP complex with a capacity of 700 MW is to be built near Dubai in the United Arab Emirates. The project developer ACWA Power expects an electricity price of 7 dollarcents per kilowatt-hour. The comparative price of European wholesale is often between 3-4 cents per kWh. Depending on irradiation conditions, records of up to 7 dollarcents per kilowatt-hour are reported in Dubai.

Dii had completely different plans in Tunisia in 2012. It worked on an integral study on renewable energies in the region. The focus was on the exchange within the Maghreb and the integral connection to Europe. In their view, a cable link between Tunisia and Italy could not be implemented in the market. Even if the option had existed, net electricity would have flowed from Italy to Tunisia and not the other way around, due to the bottlenecks in North Africa and at the same time temporary power surpluses (due to variable feed-in from renewable energies). Dii's system studies would show that the synergies through the open exchange of energy between the Maghreb countries also justify electricity connections with Egypt, Libya and the Middle East as well as Europe in the long term. At the same time, it was assumed that hydrogen that could be produced and stored in a renewable way would supplement or replace electricity grids as a means of transport to Europe in the long term.

3.2 Desert Power 2050 – The first major milestone of Dii

In June 2012, Dii published its overall report entitled *Desert Power 2050* (DP2050), which was intended to provide a guide to desert power (Dii 2012a). It was intended to supplement and update

DLR's valuable findings, which had now been available for several years. The initiative was commissioned by the Fraunhofer Institute for Systems and Innovation Research (ISI), which focuses on application-oriented developments. The Institute, based in Karlsruhe under the direction of Mario Ragwitz, sees itself as an independent pioneer for society, politics and business. Its research focuses on the investigation of scientific, economic and social conditions under which innovations develop and their effects.

The publication of the study in the summer of 2012 generated considerable media interest. A lot of figures were published and gave a visible impression of the results:

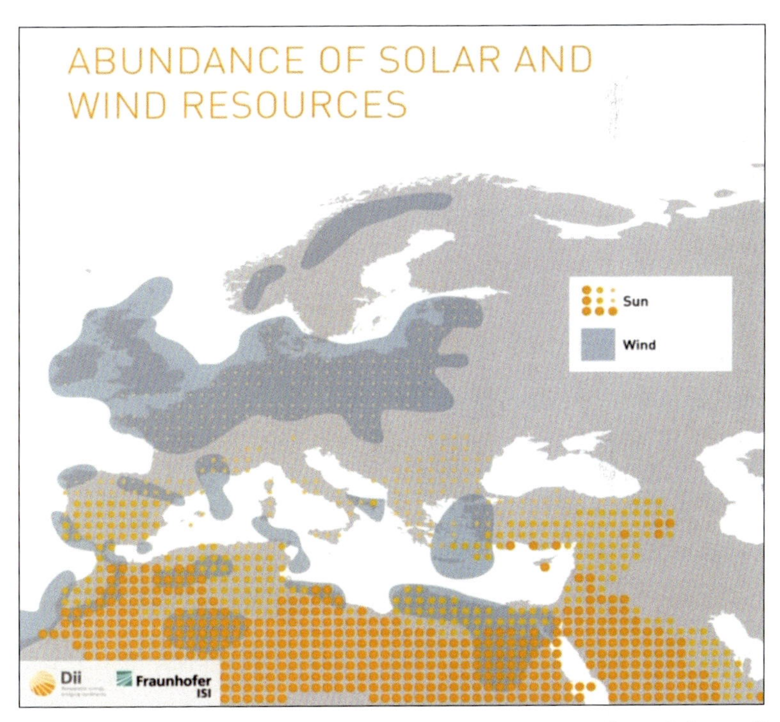

Figure 13: Abundance of solar and wind resources: MENA has excellent wind as well favourable solar resources. Northern Europe has vast on-shore and off-shore wind and hydro potential. Source: Dii

On the second page it mentions four arguments for the desert power:

- MENA and Europe need a secure, affordable and clean power supply, and that is feasible in principle.
- Supply and demand for renewable energies in the form of desert power and hydropower in MENA and wind and solar power in Europe complement each other almost all year round.
- All countries benefit from access to cheap renewable energy, emerging industries and reduced CO_2 abatement costs.
- Only through targeted, joint action by governments in Europe and MENA can this enormous potential become economically feasible by 2050.

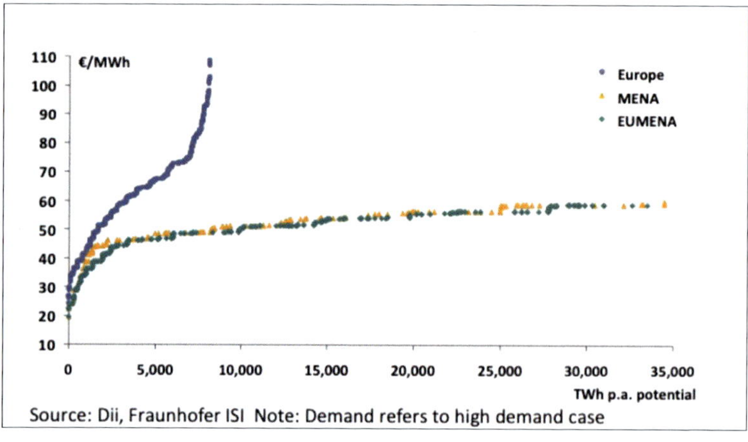

Figure 14: The basis for Dii model studies in 2012: the Solar and Wind potential in EUMENA is virtually infinite costing less than 50 € / MWh in MENA and up to 100 € / MWh in Europe: Note that in 2019 solar and wind power costs in MENA are already in the range of 30€/MWh to below 20 €/MWh.

The Kiel Institute for Global Economy subsequently found, in an evaluation of the study, that energy costs in an integrated EU-MENA energy market are significantly lower (around 33 billion euros per year) than in two regional, separate energy systems. The reason is the complementary supply and demand structure north and south of the Mediterranean Sea in all seasons (Celizalilla et al. 2014).

Desert Power 2050 further developed the thought patterns and regarded the MENA region not only as a desert power supplier for Europe. The study stood for healthy local energy markets and a free exchange of energy between countries and continents and substantiated the Desertec idea of generating renewable energies in areas with optimal resources and exporting them from there to regions with strong demand. The idea suggests that Europe would cover part of its energy through electricity imports from the deserts south of the Mediterranean. The region offers an abundance of wind and sun and also has large areas that are only sparsely populated. Europe would also be able to supply electricity to the MENA region in times of excess wind and hydropower. There was no talk of a one-way street between the regions.

According to the DP2050 proposals, electricity should be able to flow back and forth between Morocco and Norway, and Saudi Arabia and Finland in an interconnected mode. Solar energy and wind energy to the north, wind power and hydropower from Scandinavia to the south. So, everyone would be on board: the hydropower-rich Alpine countries, Great Britain, Spain, Germany, Italy and Turkey. The result of the research was to exploit synergies and no longer see Europe and the MENA region as separate regions. However, the term EU-MENA was criticized by the Mediterranean countries, which perceived it as overly offensive. DP2050 promised the north of the EU-MENA an annual cost saving of 33 billion euros with a 20 percent net energy supply through renewable energies from the MENA region. The southern part of the EU-MENA region was able to reduce its carbon dioxide emissions by 50 percent by 2050 and at the same time earn up to 63 billion euros annually through the sale of renewable elec-

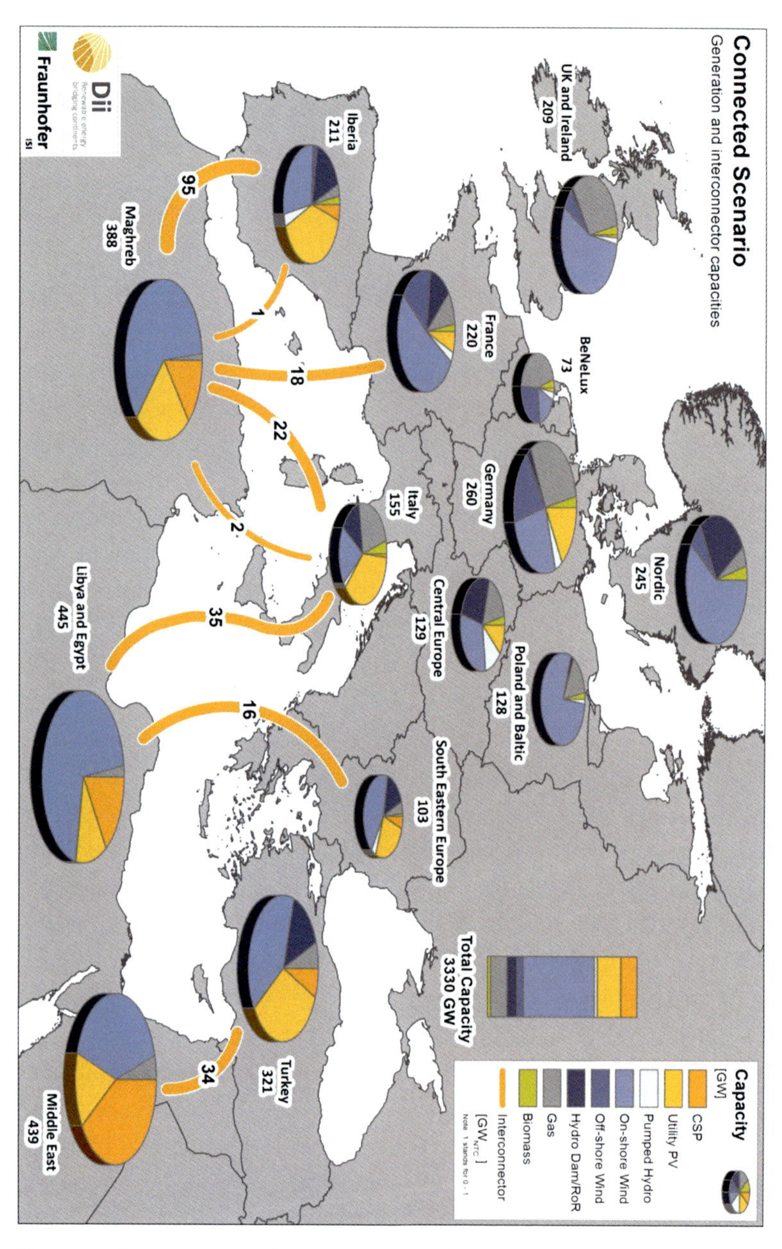

Figure 15: Generation capacity and energy exchange between Europe and MENA. Source: Dii

tricity. From Dii's point of view, the MENA region was therefore more and more a trading partner and consumer and less and less a pure producer of energy. This was a necessary development, as the study suggested, which calculated a population of 1.2 billion people for the intercontinental region under consideration. The coupling of energy markets was taken up again scientifically in the EU environment in 2015 (Van Son/Ruderer, 2015). Experts from the now 56 shareholders and partner companies of Dii were involved. The Fraunhofer Institute's ISI was able to use the already established Power ACE model, an agent-based tool for simulating electricity and emission markets, to determine the results during the period under study. At the time, the potential of solar and wind energy in the MENA region was sufficient to produce renewable energy at a cost of 50 euros per megawatt hour in 2050.

In addition to the large public presence that the study gave Dii, it now placed a central scientific result at the centre of the discussion of its now considerably grown desert power network. The energy turnaround in Europe, North Africa and western Asia would mainly be supported by decentralised photovoltaic plants (rooftop), medium-sized to large hydropower plants, solar thermal power plants (CSP) and onshore and offshore wind power plants. The study predicted a 50 percent cost reduction in electricity production from renewable sources by 2050. Only the technology for wind turbines on the mainland was considered mature in 2012 and it was forecast to reduce costs by only 20 to 30 percent. These cost expectations were still classified as very ambitious in 2012. As early as 2018, we can see that after six years, generation costs have fallen to a level that would not even have been thought possible for 2050 at the time. Among other things, because the search for new materials, manufacturing techniques and digitization have accelerated in an unprecedented global race for lowest costs. But the cost of capital is also extremely low in 2018. In a very short time, most technologies were independent of subsidies. The road to zero-emission power supply in the entire area opened up. Cost advantages also arise in desert electricity due to the more favourable climatic conditions for the generation of electricity with virtually no CO_2 emissions. In the study's scenario, a megawatt hour produced in the MENA region cost 58 euros on arrival in Europe – compared with the 73-euro megawatt hour produced in Europe.

DP2050 put forward a second important argument for creating a comprehensive intercontinental energy market: "A larger system is better able to adapt fluctuations in hydro, solar and wind power generation to flexible consumption. As a result, the number of gas-fired power plants and battery storage facilities still needed to cover peak loads is reduced. This reduction in capacity could save a further 15 euros per megawatt hour. The exchange of electricity between MENA and Europe would then be desirable and worthwhile much earlier.

According to the DP2050 study, 91 percent of the electricity mix should consist of renewable energies. Wind energy accounted for 53 percent (48 percent onshore and 5 percent offshore), solar energy for 25 percent and CSP power plants for only 16 percent. The rest of the electricity was to be generated from hydropower, biomass and geothermal energy. The dominance of solar thermal CSP power plants was no longer mentioned in DP2050. A bitter blow for the CSP industry. Nevertheless, CSP power plants, which can store energy for the night hours, played a major role.

In 2012, renewables still had a difficult position in the MENA region. At the press conference in Munich , Aglaia Wieland, who had been in charge of the studies, speculated that there were still funds available for investment in industry. However, the economic crisis in Spain and Italy meant that the investment climate in Europe was not the best, and there was also foreign policy uncertainty following the Arabellion. Many of the affected countries in the MENA region were in upheaval, some threatened to sink into chaos. German foreign policymakers in particular had reservations because they feared supply uncertainties in the strategically important energy sector. However, Germany's energy supply at that time was still heavily dependent on the supply of fossil raw materials from the Middle East. In any case, most observers agreed that more cooperation in energy transition between Europe and its neighbour, the MENA region, offered great advantages. The only thing missing was the will or the urgency to really get out of the starting blocks.

Based on Gerhard Knies' ideas, three years after the foundation of Dii the team around the managing directors Paul van Son and Aglaia Wieland succeeded in delivering the previously agreed results with DP2050 to the participating players. Finally, it could become real. Internally, Dii was not doubted. Most of the shareholders had benefited from the frequent, positive reporting in the media as well as from the scientific studies and the mutual networking. However, some companies with no strategic interest in the MENA region withdrew from Dii. At the press conference for the launch of DP2050, the gap between public perception and self-perception of Dii became very clear to non-insiders for the first time. From 2009 to 2012, Dii was regarded as a major solution for energy, climate and refugee issues and thus came under ever-increasing public expectation pressure. The study did not provide any answers to the questions about concrete construction projects and schedules for the first electricity generated, and the – unintentionally – raised expectations were not sufficiently satisfied. The public was insecure, the first Desertec ideas were rejected. Despite increasing pressure, however, Dii remained true to its line and continued to consistently position itself as a technology-open pioneer for the entire market development in the region.

3.3 It becomes real

3.3.1 A reference power plant 'against the market'?

Companies in the CSP sector came under pressure due to the drop in prices for photovoltaic modules. For their economic survival, they needed rapid sales success. In 2012, the market for this was lacking. A reference power plant initiated by Dii was to explore the possibilities of CSP technologies with the help of government subsidies.

However, the analyses of DP2050 had shifted the requirements for this: The forecast energy mix changed drastically. To the astonishment of many, the focus was no longer on CSP technologies, but on wind energy. Electricity exchange was now conceivable in both directions and the development of individual power plants was still completely open. Internal discussions within Dii showed that the realization of a single CSP reference project would have changed little. At the time of the publication of DP2050, CSP techniques lacked the corresponding capabilities. Slowly it became clear that local and international players on the market would drive these forward.

The distribution of responsibilities and influencing possibilities between the market and Dii are visualized in Figure 12. The initiative aimed in particular to address strategic thinking on individual countries, focusing mainly on the MENA region. It wanted to play the role of project developer only in the sense of a catalyst for reference power plants.

Figure 16: A parabolic trough power plant in the Moroccan desert. Source: Thomas Isenburg

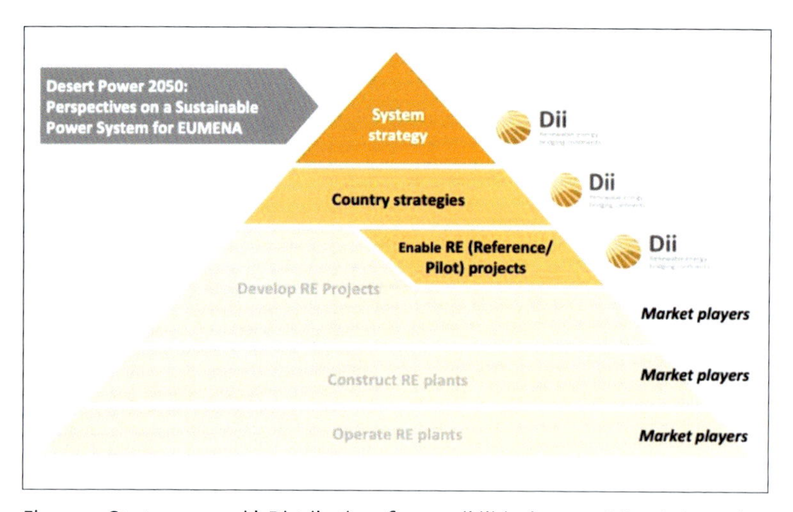

Figure 17: Strategy pyramid: Distribution of responsibilities between Dii and the market. Source: Dii

3.3.2 Hot potatoes for politics

The situation also became more difficult at the political level. Shortly after the Arab spring, parliaments in government circles were very enthusiastic about new ways of working together. This positive mood was dampened as the realities gradually became visible. Public opinion was also rather sceptical about large-scale projects such as Desertec – there were doubts as to their feasibility. As far as the import of electricity from renewable plants was concerned, there was no short-term pressure to perform at all. The market was still dictating the transport of electricity from Spain to Morocco and not, as Desertec was mistakenly understood at the time, in the opposite direction. While the average price in Spain in 2012 was 4 to 6 eurocents per kilowatt-hour, the electricity production costs in Morocco were still 7 to 8 eurocents per kilowatt-hour.

> *Although the perception was different, initially there was no grid bottleneck for export from Morocco to Spain, but there was simply no economic rationale for doing so yet.*

The governments were working on an agreement between Morocco, Spain, France and Germany to regulate the transport of electricity from renewable plants in Morocco to Europe, especially to Germany. Articles 6 and 9 of the European Renewable Energy Directive provided the legal framework. They specify the transport of electricity from North Africa to Germany and include virtual electricity supplies. Spain was fundamentally opposed when it came to transporting electricity from the deserts through the country. The Spanish government first wanted to see the bottlenecks in transport capacity to France lifted. It wouldn't negotiate major transcontinental transports before. Germany, Morocco and parts of Dii succeeded only temporarily in exerting pressure on Spain.

3.3.3 *An agreement in the margins of the Berlin conference?*

Dii planned its annual conference at the Federal Foreign Office in Berlin for November 2012. The political leaders had agreed to be present, among them Foreign Minister Guido Westerwelle, Economics Minister Philipp Rösler, and EU Energy Commissioner Günther Oettinger. Arab ministers and state secretaries were also on the guest list.

On the sidelines, diplomatic efforts were made to find a diplomatic solution to the controversy with Spain, but the time did not yet seem ripe. Morocco and Spain, on the one hand, and Spain and the other European countries, on the other, have long exchanged electricity despite limited cross-border transmission capacities. Morocco, for example, structurally purchased electricity from Spain because of its energy deficit via the power lines under the Strait of Gibraltar. The existing infrastructure would have been sufficient to transport electricity from Moroccan renewables to Europe if there had been interest. Through ONEE, Morocco also traded on the Spanish power exchange. The intention to explicitly regulate the export of electricity from renewable sources, i.e. mainly CSP electricity, from Morocco to Spain and then via France to Germany, did not fit into the market model of European electricity suppliers. Combining the implementation of renewable energies in the MENA region with green certificates, as proposed by Dii, would have been an elegant way to resolve the dilemma of electricity promotion and market requirements. However, some German government agencies rejected this approach. A breakthrough in intercontinental electricity exchange – combined with the promotion of renewable energies in Maghreb – was not in sight.

3.4 *Old and new players*

3.4.1 *Siemens and Bosch are leaving*

At the end of 2012, further dark clouds appeared in the sky of Dii, which had been blue for most of the time. Unlike most other shareholders, Siemens did not opt for a two-year extension of the Dii mission, and Bosch also left the group of companies (dpa 13.12.2012). The companies listed in the DAX no longer extended their contracts, partly because they wanted to reduce their involvement in the solar industry. The Bosch Group, which had previously been active in the production of photovoltaic modules, gave up this division altogether. In 2009, Siemens had bought Solel and invested in solar thermal technology, which turned out to be a mistake. On the one hand, the technology was regarded as expensive, on the other hand, in countries such as Spain, the solar subsidy was dropped. Moreover, the Israeli company did not have an easy position in the Arab market. To this end, the Munich-based technology group continued to operate its wind energy business with great intensity. Siemens, in particular, had a reputation in Germany for strongly promoting the idea of desert power; in Switzerland, the same was true for ABB.

Due to low-cost mass production in Asia and the resulting massive drop in prices for photovoltaic modules, the German solar industry came under increasing pressure, as already explained (Baumer/Kepler 2017). At the same time, a trade dispute between European and American manufacturers intensified. In particular, German companies such as Solar Millennium and Q-Cells had benefited from the generous EEG subsidies. The market for solar modules dropped by 60 percent in 2013 and by 45 percent in the first half of 2014 as a result of sharp cuts in feed-in tariffs. Siemens justified its decision to turn away from solar energy by concentrating on business with gas turbines, hydroelectric power stations and wind turbines. These developments would not remain without influence on Dii.

The hopes of those who had started together had been great. They bet on financial profit and a win-win situation. The market shifts and the results of DP2050 had now changed the situation. The interests of the shareholders did not always coincide with the new findings. The managing directors of Dii were faced with the difficult task of reconciling the shareholder interests of an increasing number of partners and the TuNur activities of the Desertec Foundation with the study results of Dii. Within the team, this extremely complex task was to become the cause of major tensions and far-reaching conflicts, which in turn had an impact on the renowned companies and banks. Confusions arose, also in the external picture, and the companies involved no longer identified completely with Dii.

3.4.2 *Drilling thick boards*

In the meantime, Desertec had become a laborious business, was the opinion in the circle of shareholders. However, some companies felt that they should happy about everyone who was willing to get involved. And international interest actually grew. Van Son brokered contact with the Chinese transmission system operator State Grid Corporation of China (SGCC). With 1.7 million employees, it is one of the largest companies in the world. SGCC advocates intercontinental electricity exchange, among other things, and made proposals for a global energy network (Uken 06.11.2012). Chief strategist Aglaia Wieland, who had already been appointed second managing director in February 2012, was opposed to the Chinese and tried to prevent their admission to Dii. The American company First Solar and the rapidly rising project developer ACWA Power from Saudi Arabia also became new shareholders. The initiative became more international and diversified, which increased the pressure on Dii to succeed from within.

Arab Spring has changed the political map of the southern shores of the Mediterranean. The conditions in Libya and Syria had complex international repercussions. Behind the scenes of Dii it has been rumbling for some time. The leadership and management level, which was still dominated by German culture, lacked a culture of open discourse, and points of view were hardly discussed openly. In particular Wieland and some of her comrades-in-arms continue to push for a first reference project of Dii in Morocco, without seeking connections with Morocco's efforts to implement its own CSP power plants. The German government should provide funds for the expensive, electricity exporting CSP power plant. Van Son was critical of such plans. He already considered the planned CSP projects of MASEN in Ouarzazarte, Morocco, to be excellent references for the market. Since the exchange of electricity between Morocco and Spain had been functioning perfectly for many years and

Morocco still had a large energy deficit for the time being, he saw no reason for a reference power plant of its own that would not contribute to Morocco's energy supply.

3.4.3 *The big Dii Conference in Berlin*

To welcome the participants at the annual conference of Dii at the Federal Foreign Office in November, the participants received the mock-up of an EU-MENA passport as an indication of the growing together of the regions of Europe and MENA. The shareholders of Dii sent prominent representatives such as Thorsten Jeworrek of Munich Re, Giuseppe Vita, Chairman of Unicredito, Caio Koch-Weser of Deutsche Bank and Hans Bünting, a pillar of RWE-Innogy. Mustapha Bakkoury, CEO of MASEN, Mouldi Miled, CEO of the Desertec University Network, Jonathan Walters of the World Bank, Kevin Sara, CEO of TuNur and about 400 other guests also came to the German capital to exchange ideas in crowded conference rooms. Young Arab journalists who had received a travel scholarship from Dii to be able to report in their home country were also present.

Figure 18: Günther Oettinger at the Dii Conference in Berlin. Source: Thomas Isenburg

Shortly before the conference, Spain withdrew from the negotiations on electricity transport, and a Memorandum of Understanding was not concluded. A Moroccan Secretary of State provided the public with the relevant information. The government representatives of Morocco, France, Italy, Malta and Luxembourg had all come to Berlin, Spain had not sent any representatives. Ministers Westerwelle and Rössler were represented contrary to earlier announcements.

The European Union Commissioner for Energy, Günther Oettinger, a proponent of more renewable energies in the deserts also with regard to the European electricity supply, pleaded in Berlin for the expansion of national grid structures to transmission grids in order to integrate the whole Mediterranean area. However, he also noted that the EU lacked the necessary money for this. MASEN chief Bakkoury stressed at the joint press conference with Van Son that a strategically important issue such as energy export needs sufficient time. Van Son added, "Patience and humour are two camels walking through the desert."

At that time, every CSP power plant in Morocco would have served the Moroccan market, as we know today. All countries in Europe and MENA were primarily concerned with their own energy concerns. Spain continued to export electricity to Morocco for the time being. The German energy market was busy with the rapid growth of renewable energies in its own country –a Herculean task. However, the discussion on cross-border electricity connections in Europe had received a new impetus.

The work of Dii increased the interest of the German energy groups E.ON and RWE. They cautiously thought about a commitment in the MENA region. At the annual conference, E.ON had set up the model of a CSP power plant in order to demonstrate its interest in working with some Dii partners on a CSP project in Morocco. RWE preferred to develop its own PV or wind project with a capacity of around 50 megawatts, if possible also in Morocco. To this end, the Essen-based energy professionals wanted to put together a consortium of Moroccan and international companies by the end of 2012. Both energy managers gave the supply of electricity to Europe and the concept of green certificates opportunities if the profitability of the projects improved. Dii supported the projects at the political level through talks between the EU, Germany and Morocco.

3.4.4 *Morocco, the fast mover*

Morocco continued to actively develop wind and CSP turbines. The plan to expand wind turbines with a capacity of 2,000 megawatts was managed by the ADEREE energy office under the leadership of Said Mouline. The Société d'Investissements Energétiques under Ahmad Baroudi was prominently featured in the news with plans for PV systems on all mosques in Morocco. At MASEN, a group of young people, most of whom had studied abroad, gradually implemented the Moroccan solar plan. They prepared tenders for power plant complexes in the order of 100 to 200 megawatts. Initially, the Moroccans concentrated primarily on solar thermal CSP power plants. For implementation they received loans from development banks such as the German KfW Bank, the European Investment Bank and the French and African Development Banks. The EU and the Clean Technology Fund of the World Bank also contributed. Bakkoury was familiar with the capital flows of the international financial world and skilfully initiated development projects for renewables.

ACWA Power, the power plant developer based in Riyadh and Dubai, won the tender. Its director is Paddy Padmanathan, a dynamic managing director and visionary with extensive international experience, whom Van Son met in 2011. Padmanathan, who studied in England, has managed the company since it was founded. ACWA Power quickly turned out to be the driving force in Dii. The Saudi company was engaged in the development of conventional and renewable power plant complexes and desalination plants. The turnover of 200 million euros is generated by 3,000 employees from all over the world. Many projects were realised in North Africa and the Middle East, but also in South Africa, Bulgaria, Turkey and China. With this portfolio, it fitted almost perfectly into the concept of Dii. Padmanathan initially built fossil-fuel power plants, but with ACWA Power he wanted to focus increasingly on renewable energies. The company possessed the resources to build large power plants; by participating in Dii, it should be able to further strengthen its global business.

Padmanathan never missed a shareholders' meeting and his work soon bore fruit. An excellent businessman, he quickly got to grips with the matter, oversaw the dilemma of Dii, and knew how to use this knowledge. ACWA Power also became active in Germany by taking over the insolvent manufacturer of solar mirrors Flabeg, from Furth in the Bavarian Forest. The company succeeded in becoming the market leader in the tender procedures in the MENA region. German companies, such as Schott and Siemens, and the Swiss company ABB, which has its roots in Germany, acted as suppliers to the Arab power plant developer.

Expert voice: Paddy Padmanathan (CEO ACWA Power)

Paddy Padmanathan. Source: ACWA Power

Interview with Paddy Padmanathan, CEO ACWA Power

ACWA power was founded ten years ago as project developer for water treatment and electricity sector. Building on our renewable energy activities in 2011, we came across Dii and its mission to create a renewable energy market in the MENA region. They promoted our work with international, local, political and industrial stakeholders. It enabled us to collaborate with international, local, political and industrial representatives. The focus of ACWA is climate protection, but also the creation of new industries, investments in the region and jobs creation. Indeed, the transfer of knowledge and the know-how is an important factor. Policy makers and lobby groups need well researched case studies from the private sector. Therefore, we wanted to show that renewable energies are feasible with concrete projects.

ACWA enthusiastically joined Dii and our intention was not to generate new business at short notice. Our first interest was to support countries in their investments on renewable energies. We knew, those investments are not free of risk and they are also commercially unattractive. Prosperity and stable development in the region are therefore the most important pre-requisites for sustainable business prospects. We believe in this and support it. Dii has produced valuable studies and demolished myths and misunderstandings. They caught the attention of the political makers on desert power. The rapid progress puts the costs for PV and wind energy down. Today, they are competitive with fossil resources. The implementation of renewable energies is now progressing at a faster rate. However, there is still a lot of work ahead in order to persuade the policy makers of the market value for renewable energies.

The Energy supply without CO_2 emissions is absolutely possible. Dii's network is very valuable to ACWA, which is why we always have extended the originally agreed mandate to this day.

Saudi Arabia is a worldwide energy supplier for more than 60 years. The 33 Million population needs a lot of energy for air conditioning and desalination. The country's energy policy is still far from being future oriented. The costs for electricity and desalinated water are subsidized. The production costs for fossil raw materials are low and the country still provides its citizens with energy at a very low cost. This promotes the inefficient use of energy.

The implementation of renewable energies has been in early days difficult because of the high prices, yet this has changed. The Kingdom is now re-evaluating its oil revenues and adding renewable energies to its energy mix. This will also contribute to sustain the country's position as energy export nation. Finally, Saudi Arabia noticed that the implementation of renewable energy has created new jobs for young people.

3.4.5 *The extension of the Dii shareholders' agreements*

At the end of 2012, when the contractually agreed result had been delivered with the DP2050 study, the contracts with Dii were due for renewal (Dii 2012b). Initially, further cooperation with the shareholders was agreed for two years. Dii had 21 shareholders and 35 associated partners at that time – after having started in 2009 with 12 players. But the companies in the CSP industry that had high hopes for Desertec were disappointed. Following the construction of power plants, particularly in the USA and Spain, further orders failed to materialize and the industry's existence was threatened. The displeasure about it did not pass Dii without a trace. The energy mix planned in 2006 with CSP technologies as the supporting pillar – the original Desertec idea – was called into question by the massive fall in the price of PV modules in previous years. At the end of 2012, the price was around 0.08 to 0.10 euros per kilowatt-hour for photovoltaic systems on open spaces. CSP power plants at similar locations were able to produce electricity for 0.18 to 0.23 euros per kilowatt-hour and store it for several hours. Worldwide, a capacity of 2 gigawatts was installed at CSP power plants. However, this was not enough to achieve economies of scale for a significant reduction in costs. For the power plant complex in Ouazazarte, a price of 0.14 euros per kilowatt-hour was forecast at this time. DLR adhered to the concept of solar thermal power plants and direct lines to Germany. She was unimpressed by the drop in the price of PV modules and put the potential energy storage of this type of power plant in the field. For the major projects implemented in southern Spain and the USA, most of the suppliers came from Germany – in Ouarzazate things should be different. ACWA Power worked with the Spanish construction company Sener and various subcontractors. The Spaniards in particular cultivated a close relationship with their North African neighbours and felt at home due to their centuries of common history. German technology was also used, such as the receiver rods in the center of the parabolic troughs of Schott Renewables.

In the meantime, intensive negotiations continued in Germany. Most companies wanted to stay in Dii – for different reasons. Munich Re remained loyal to its work with Dii. It continued to look for a way to an emission-free energy supply because it feared horrendous damage from climate change. The development of the huge springs in the deserts was still obvious for them, their attitude had not changed since the foundation of Dii. ABB, represented by Kreusel, also appreciated the work of Dii. International companies, such as First Solar, ACWA Power, Red Electrica, Terna Energy (Greece), Nareva (Morocco) and Saint Gobain (France) remained as shareholders, with a number of associated partners. Van Son saw the work of Dii confirmed by this. However, Dii was not as well received in the press as it was at the beginning. For the journalists, there were no headlines, such as on the construction of large power plants. This was interpreted as a threat of failure. Undisturbed by this, the industry initiative pursued the agreed mandate in a resilient manner. The team was supported by several external experts: Gerhard Hofmann, known from the Berlin media and political scene, the company Hengeler Mueller, which was on board as a pro bono legal consultant, Silvia Kreihbiel from Deutsche Bank for financial questions and Kirsten Westphal from the Stiftung Wissenschaft und Politik (SWP) in Berlin are just a few examples. An Advisory Board headed by Professor Hans Müller-Steinhagen of DLR, which included former Tunisian Energy Minister Abdelaziz Rassâa and Driss Benhima, CEO of Air du Maroc, advised the Executive Board intensively. Especially the political and legal framework conditions in the individual MENA countries are complex and often not transparent. They raised countless questions. Nothing seemed impossible, but German circles still lacked the necessary confidence.

3.5 Turbulences

3.5.1 Forerunners of the dispute

At the beginning of 2013 it became clear that the desert countries would benefit from electricity production along the entire value chain. This included electricity exchanges between countries and continents. However, they were only able to become exporters thanks to the generally good production conditions. In the coming years, Morocco's energy supply would still depend largely on the primary energy sources of hard coal and natural gas. The country, rich in renewable resources, would have to continue importing coal from South Africa and gas from Algeria at high prices. The Moroccan wind and solar plan was a step in the right direction. But it would take a long time before the energy mix was sufficiently emission-free and deliveries to Europe seemed feasible. So, it made no sense to talk about transporting electricity from Morocco to Germany in the short and medium term. At that time Van Son saw the previously envisaged Dii reference power plant only as a virtual project, as an example for the market, and advised that Desertec should no longer be presented as a one-dimensional electricity post-Europe project. Internally, he announced that Sawian (Arabic: cohesion), the physical reference project for transporting electricity to Europe, was dead. This led to tensions within Dii, as Abengoa and Schott Solar, two companies that were shareholders, were desperately looking for orders for

CSP power plants. MASEN would probably have been open to any project that would have been lavishly rewarded with prices of around 25 eurocents per kilowatt-hour. In the team of Dii polar opposites now formed around Wieland and Van Son – the prelude to violent internal arguments. Moreover, the sting of the colonial past in Morocco and the MENA countries was often still deep. Over a hundred years later, the countries increasingly understood that they had to take the initiative themselves. However, the individual cities, countries and emirates felt different pressures to act and worked under different political and legal frameworks. For social reasons, there were still high subsidies for electrical energy. This combination created a certain lethargy in the markets. The established fossil interests had a strong lobby and the implementation of renewable energies continued to be difficult. The Hamburg Institute of International Economics declared in spring that the Desertec project had lost its appeal (Wolf 2013).

Nevertheless, in various MENA countries, large-scale renewable projects for domestic supply were on the political agenda, which had already been reported in Morocco and the United Arab Emirates. In 2011, the CARE (Center for Atomic and Renewable Energy) in Saudi Arabia also made big announcements: between 2012 and 2032, the Kingdom wanted to invest 110 billion US dollars in renewables (Wetzel 19.02.2013): in solar thermal power plants with a total capacity of 25,000 megawatts as well as photovoltaic systems with 16,000 megawatts. For size comparison: in Germany there was an installed photovoltaic capacity of 30,000 megawatts at the time. Competition for renewable energy capacity in the region was open. In the meantime, there was even a renewable energy plan in Algeria, a success on which Dii had collaborated in a study with Sonelgaz.

In 2013, the foundation stone was laid for the 160 megawatt parabolic trough power plant Noor I in Ouarzazarte, Morocco.

King Mohammed VI attended the ceremony. The Noor complex consisted of four solar power plants and would have an output of 570 megawatts. The World Bank estimated the total cost of the project at USD 2.677 billion (as of 2015), of which Germany – i.e. the BMZ (Federal Ministry for Economic Cooperation and Development), the BMUB (Federal Ministry for the Environment, Nature Conservation and Nuclear Safety) and KfW – assumed EUR 829 million. Accordingly, the German public was interested in the power plant and also in transporting electricity to Europe. Under the management of ACWA Power, the Spanish company Sener began levelling the 12 x 12 kilometre site. Since the CSP plant was still designed for water cooling, a building for the water supply from the nearby dam was erected in one of the first project steps.

Figure 19: Start of construction in Ouarzazate. Source: Thomas Isenburg

In Ouarzazate, at the gateway to the Sahara, there was still little of the bustle around the large solar power complex to be felt. Only a few residents suspected that technical history was being written in their neighbourhood.

3.5.2 *Team dispute and desert power: getting started*

Meanwhile, in the Dii team in Munich, a good 3,000 kilometres away, things started to reach boiling point. Van Son and Wieland argued about the future strategy: Van Son wanted to abandon the idea of a reference power plant with the goal of the rapid export of desert electricity to Europe for the time being. Wieland continued to pursue this goal unswervingly. This was followed by months of intensive and controversial discussions, which unfortunately also found their way into the public eye.

In June 2013, the second Dii study, *Desert Power: Getting Started* (Dii 2013), was presented to the public at the Sustainable Energy Week of the European Commission in Brussels. It is a continuation of the first Dii study, *Desert Power* 2050. Media interest had already declined noticeably and was now more focused on the heated conflicts within Dii. *Desert Power: Getting Started* describes regulatory and financial aspects for a successful development of renewable energy in the MENA countries for the years 2020, 2030, 2040 and 2050. The study should be seen as a kind of guideline for renewable energy in the MENA region.

It showed which obstacles still had to be overcome on the way there. In the MENA region under consideration, there was a clear cost advantage for electricity production from renewable energies compared with Europe. Oil and gas importers such as Jordan, Syria and Egypt could save their budgets by implementing renewable energies. Oil and gas exporters like Saudi Arabia. Algeria, Qatar, the United Arab Emirates and Libya could increase their income as more oil and gas would be available for sale with the help of renewable energies. To ensure a secure energy supply, local and transnational power grids should be expanded everywhere. The study saw grid bottlenecks in transport to Europe in Spain and Italy. Up to 50 gigawatts of capacity could be installed in the entire MENA region by 2020. For comparison, the

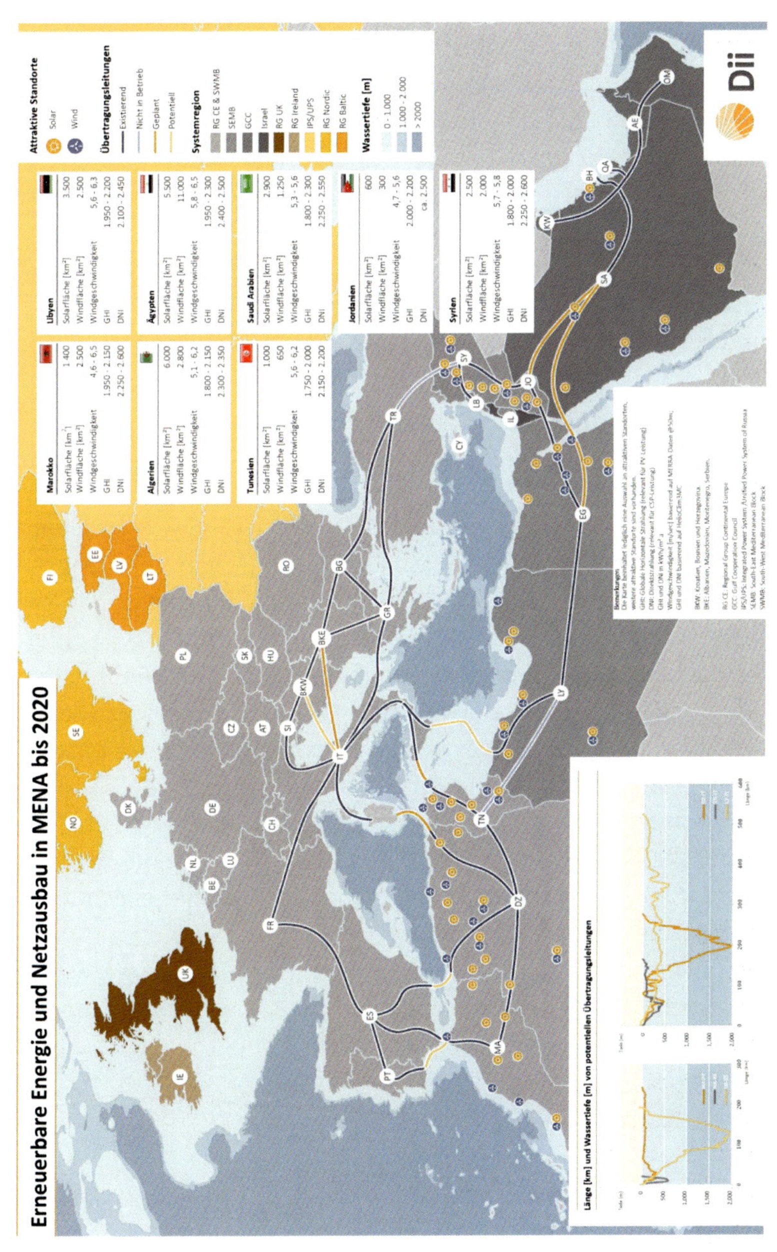

Figure 20: Forecasts for grid expansion in MENA 2020 in Desert Power: Getting started. Source: Dii

Moroccan solar plan, which is considered ambitious, assumed a capacity of 6 gigawatts to be installed during the same period. If implemented according to plan, almost all countries would have had to push ahead with similar projects. Here, too, a weakness of the original Desertec concept became apparent. The plans were devised by European minds and missed social, economic and societal realities. The strength of Dii was to set social processes in motion. In all the inevitable conflicts it managed to bring international industry and the right people together and initiate processes.

3.5.3 *The media bomb explodes*

At the end of June 2013, a report in the *Süddeutsche Zeitung* entitled 'Schatten über Sawian' (Shadows over Sawian) brought the dissent between Van Son and Wieland out into the public eye (Basler 27.06.2013). This was preceded by fierce internal disputes. A reaction was not long in coming: the Desertec Foundation announced their withdrawal from Dii three days later (Basler 30.6.2013). Gerhard Knies regretted the situation but encouraged Dii to continue with the development of renewable energy. The communication between the Desertec Foundation and the other shareholders of Dii was difficult. For the first time, an NGO and large industrial groups had come together in the legal form of a limited liability company (GmbH). A wasted opportunity. Following the withdrawal of the Foundation, both were able to make a clearly separate contribution to the common goal of an emission-free energy supply and the strengthening of desert countries.

3.5.4 *The separation*

The conflict between Wieland and Van Son escalated at the Dii shareholders' meeting in Seville, Spain at the beginning of

June. It was about the extension of Van Sons contract. Wieland, completely surprisingly for many participants, strove for sole management, as a covert formulation in an e-mail to some shareholders on the eve of the meeting suggested. Critics of Van Sons interpreted his scepticism about an imminent export of electricity to Europe as an economic interest. So did the *Süddeutsche Zeitung*, which mentioned Van Sons proximity to RWE – he was once the manager of the RWE subsidiary Essent (Basler 27.06.2013). Van Son defended himself: "Electricity exports will take place when the time is ripe. The short-term aspects of the Desertec vision are still strongly influenced by European wishful thinking." He also warned against the exaggerated public expectations created by the media hype. In Seville, his contract extension was confirmed as planned. The shareholders appreciated his in-depth knowledge of renewable energies and electricity grids. In the meantime, the Dutchman was excellently networked in the MENA region and knew the connections between the international energy systems very well. Competences that were existentially necessary for a project like Desertec. In addition, the shareholders took positive note of the manager's latest insights.

The public battle for direction increasingly paralyzed Dii. In a letter to the company representatives, the Dutchman criticised that since the last shareholders' meeting it had become "almost impossible" to make decisions. The dispute was characterized by personal vanities and different business interests. The prominence and the considerable importance of the project now came back like a boomerang. The conflict over a reasonable strategy ended with the release of the second managing director and head of strategy. This decision was supported by a large majority, with no dissenting votes from the shareholders. Wieland had lost the power struggle. A decision was made at a crisis summit at Frankfurt Airport (Hoffmann 09.07.2013).

3.5.5 *The show will go on*

Dii's initial image as a catalyst for sustainable energy supply in the MENA region and in Europe was tarnished. Internally, the team now led solely by Van Son had to redefine its location. The core shareholders stood behind the Dutchman. In October, the annual conference of Dii took place in Rabat. There, many discussions revolved around the large Ouarzazate construction project. Among others the Moroccan Minister for Energy Abdelkader Amara, Mustapha Bakkoury, delegations from Algeria, Tunisia, Egypt, China, Germany, Italy and Spain as well as the shareholders and partners of Dii had come.

Figure 21: The Dii Conference in Rabat. Source: Thomas Isenburg

The CEO of Saudi project developer ACWA Power, Paddy Padmanathan, also gave a presentation. Padmanathan and other international stakeholders are using the power and knowledge of Dii, which has long ceased to be a German initiative; rather, the trend towards internationalisation has continued. Dii had become an open technology platform for the entire market.

3.6 The new search process

After the management dispute, which had been broadcast to the entire group, and the publication of the new studies, Dii was in an open search process. It operated more under the radar now and continued to work on content. The *Getting Started* study was undisputed and contained valuable information on the interrelationships between Europe's energy systems and the MENA region. In 2009, PV systems were not considered to be competitive with CSP power plants. Also, the very large possibilities of wind turbines were completely underestimated at that time. This had changed in the meantime. Chinese companies entered the market with their PV products. The wind market was mainly served by German, Spanish and Danish companies.

People in Europe were busy with the energy system transformation. Large lignite, coal and nuclear installations became under pressure due to their large CO_2 production. Renewable electricity, generated centralized or decentralized, cannot easily replace such systems. Their production is volatile. Flexible consumption, storage and exchange with neighbouring networks could, however, compensate for decreasing controllability over time. The development of efficient storage technology has moved to the top of the research policy agenda. The Dii project should have a visible effect on the energy supply in (Northern) Europe only after a few decades. There was a lack of suitable pictures to inspire the rather pragmatic top managers who were represented in Dii even more. Even the government representatives did not do this easily. At the end of 2014, the contracts concluded for a limited period in autumn 2009 would expire. From within Dii, the idea was developed to make it a smaller, permanent institution.

After the Arab spring, a whole series of NGOs had developed in the Arab world, dedicated to environmental issues. More and more regional and international institutions became interested in desert energy, which, however, received less and less attention in European politics. To this end, the southern shore of the Mediterranean was opened up to renewable energies. Here the work of Dii bore fruit and interest in environmental issues grew, especially among young people.

3.7 Dark clouds in the skies of international politics

In the MENA region there were fierce political conflicts. In Syria, as in Libya, the protests of the people against their totalitarian rulers Assad and Gaddafi led to devastating civil wars. The major international powers interfered, also because they considered the markets for fossil raw materials to be in danger. On the European side, France intervened massively in Libya, which is rich in oil. The country had the highest per capita income on the African continent through the sale of its oil resources. Egypt, too, was boiling. Peaceful protests in Tahrir Square had led to the deposition of President Mubarak. His successor Mursi was chosen in free elections. In 2013 the masses rose again on Tahrir Square. It came to a military coup and general Al-Sisi, trained in the USA, took over the government and was later confirmed in elections. The Egyptians longed for peace and stability. In the Gulf region, the interest of Saudi Arabia and the United Arab Emirates in renewable energies increased. They want to use their favourable climatic conditions for the production of renewable electricity to use fossil resources for export. In Tunisia and Morocco, too, the development after the Arab spring was positive. The *Quartet du Dialogue National* received the Nobel Peace Prize in Tunisia for its decisive contribution to building a pluralistic democracy. Morocco's commitment to renewable energy has been highly praised internationally and Marrakech hosted the COP22 Climate Change Conference in 2016.

Figure 22: Liangzhong Yao, Senior Executive, State Grid of China, CEPRI at the signing of the contract with Paul van Son. Source: Dii

An important prerequisite for the Desertec project was a policy that promotes détente, because the vision was based on intensive cooperation between Europe and MENA as a connected energy region. Where stability is evident, where the geostrategic position is interesting and where people are motivated, there is an investment-friendly climate. Morocco and Tunisia in particular met these criteria. In Morocco, free trade zones developed around Casablanca and in Tangier, located on the Strait of Gibraltar, which were used by the automobile and aviation industries to set up production facilities.

Dii adapted to the changed situation. At the end of the year, it expanded its shareholder base to include the CEPRI research institute of the Chinese network operator State Grid Corporation of China (SGCC). A further step towards the internationalisation of the industrial consortium. New players from Asia are stepping onto the scene with structured considerations for the development of a supergrid. Among others, the People's Republic of China, headed by State Grid CEO Dr. Liu Zenya, plans to build a global power grid with international partners. The Chinese electricity grid operator SGCC is developing the plans for this. A driver for this is Chinese President Xi Jinping, the supreme leader of his country. He is committed to the global exchange of energy. These spatial thoughts should connect countries and continents with 'Ultra-High Voltage DC grids' and local 'smart grids' as comprehensively as possible and thus reduce the overall costs for the global power supply. To this end, the Global Energy Interconnection Development & Cooperation Organization (GEIDCO) was established in March 2016. The initiative for this was taken by the State Grid Corporation of China (SGCC), a shareholder of Dii. The organisation now has 260 members from 22 countries. A strategy is being developed for a global supergrid by 2050. Based on renewable energies, this global power grid is to be expanded step by step across countries, continents and finally intercontinental. According to GEIDCO, by 2050 the new network infrastructure could absorb around 80 percent of primary energy from renewable sources. In addition, the aim is to eliminate energy poverty in countries in Asia, Africa and Latin

America. The cost of this is estimated at USD 50 trillion. Much is reminiscent of Gerhard Knies' proposals, but the dimensions are even more gigantic. The Desertec idea fit with these plans, and the Chinese became an important and respected partner of Dii.

Expert Voice from Dr. Xiang Zhang Lei, Director General Europe of State Grid Corperation of China & Director General of GEIDCO, Europe

Dr. Xiang Zhang Lei. Source: privat

In China we have followed the Desertec vision with great interest from the beginning. The industry network of Dii has drawn our attention as we share the objectives of making effective use of quasi unlimited renewable energy sources for the people in the region and to connect large (and weather dependent) energy sources with demand centers at distances of several thousands of kilometers across national borders. We were aware that we could initially learn from Dii and Dii from us, as China has built up an extended Ultra High DC Voltage Grid (UHVDC), connecting large solar, wind and hydro capacities with the main demand centers in the East and South of the country.
In 2012 we have met Paul van Son in Berlin and we discussed with him the adoption of State Grid's subsidiary CEPRI as a shareholder of Dii. This was effectuated in 2013 to great mutual satisfaction.

Our staff has ever since worked with Dii and its industrial partners, exchanging views and practical information, among others on the ten UHVDC lines that have been put into operation in China and three lines under construction. The present '1100 KV Jiquan project' is scheduled to resume

power transmission at the bipolar low end in the summer of 2019, including four converters. The converter station of 'Zhangbei Rouzhi Project' has finished its civil construction and started its electrical installation. It is planned that the Zhangbei-Beijing will be put into full operation on mid 2020. In 2016 the Global Energy Interconnection Development Cooperation Organisation (GEIDCO) has been founded on the initiative of State Grid to promote and develop power grid connections between countries and continents across the globe in order to allow the secure and economic adoption of major amounts of renewable energies. Dii became soon one of the first members of GEIDCO.

In the meantime GEIDCO has grown to an international organisation with over 500 members from more than 80 countries. We heartily recommend GEIDCO's flagship publications on global energy interconnection, among others zooming in on key subjects such as Africa, North East and South East Asia, Environmental Protection and the Implementation of the Paris Agreement (ref. www.geidco.org). In cooperation with Dii, SGCC/CEPRI and the Gulf Coordination Council Interconnection Authority (GCCIA) GEIDCO has contributed in 2018 to an exploratory study on the adoption of PV in the grid and a connection between GCC and India. This is only one of several examples of a fruitful cooperation with Dii and other international stakeholders. We are proud about our joint achievements and we looking forward to more joint activities in the future.

4. Disillusionment and new perspectives (2014-2018)

The first five years of Dii, from 2009 to 2013, were turbulent. Results were quickly achieved which still have substantial significance in 2018, i.e. about ten years after the foundation. One could hardly have expected that the implementation of the largest infrastructure project of its time would become easy. It was designed for 40 years, 60 countries and three continents were involved. Before Desertec, awareness of the importance of renewable energies in the desert was not particularly pronounced in the region itself, nor in Europe, the USA, Japan or China; the ideas were primarily of academic interest. The great response to the announcement of Dii 2009 in Germany and far beyond had therefore not surprised anyone.

A few years later, none of the images that were effective in the media at the time coincided with the realities of this great energy transition. The developments did not follow the scenarios of the research institutes. However, the technologies for efficient energy conversion alone developed much faster than the optimistic scenarios had predicted. In less than ten years, the costs of photovoltaics have fallen by a factor of more than 10. In 2009, solar thermal energy was still used as the basis for energy supply. A performance that is outdated today. No one should have the illusion of being able to produce a fixed timetable for the next 40 years. What was announced as a project back then was and is nothing more than a movement, a search process, through trial and error, but amazingly effective and goal oriented. It is not only about more renewables, but also about less fossil energies in the energy mix, with absolute profitability and preferably without state subsidies. A process that should lead to an emission-free future by 2050 at the latest.

4.1 Desert Power: *Getting Connected*

Under the direction of Power Grids Manager Philipp Godron, Dii continued to work on the Getting Connected network study, its remaining task. Throughout Europe, in most parts of North Africa and in the Middle East, there are transport networks which are often linked across borders with the large interconnected networks. They are mainly operated with alternating current (HVAC). The power plants are mostly located near the consumption centres. As a result, there is only a small transport requirement for electrical energy. Problems such as the failure of a power plant can be compensated by the structure of the electricity grid. The interconnected networks are also used for electricity exchange in regions with an open market for electricity trading. Where the energy sources are far away from the consumption centres, or the distance between the markets is very large, high-voltage direct current connections (HVDC) are worthwhile. If there are problems with essential connections between power plants or with a cable connection under the Mediterranean, this has consequences for the load flows. If possible, the networks are therefore dimensioned redundantly, so that local disruptions do not lead to an interruption of the power supply.

The third major Dii study therefore dealt with the expansion of the electricity grid between the continents and countries for the use of desert electricity (Dii 2014). At the time of the work on *Getting Connected*, there were already a handful of high-voltage DC transmissions between England, mainland Europe and Scandinavia. Their planning and expansion took several years and was capital-intensive. Other examples are the three-phase connection between Morocco and Spain, the Sacoi connection from Italy via Sardinia to Corsica or the privately operated connection from Italy to Montenegro, all HVDC. As mentioned above, cable connections in the Mediterranean are difficult to install, due to the depth and rocky subsoil. Various studies have analysed these conditions in detail.

The study *Desert Power: Getting Connected* (Dii 2014), presented in the summer of 2014, assumed investments of 60 billion euros in network infrastructure in 2030 and about 550 billion euros in 2050. It differentiated between purely national, international and intercontinental connections. Connections between the MENA region and Europe are estimated to devour about 15-20 percent of the total investment of energy transition. Initially, connections between Italy, Tunisia and Algeria were considered.

The study rightly mentioned that it was difficult to look into the future. Economic, political, technical and social developments were not accurately predictable. For example, it was not known how alternatives such as hydrogen and other synthetic green media would develop and how strong the political and social pressure would become. The question also arose as to whether the political agreements on climate protection and emission-free energy would be implemented. These considerations were accompanied in the background by a group consisting of the Italian companies Enel, Terna and the organisation Res4Med. They wanted to bring Dii to Italy and focused their attention on the Dii conference in Rome.

4.2 Rethinking

A great disillusionment arose among many participants. Although the studies had delivered reliable results, short-term project ideas had come to nothing, even if important lessons could be drawn from them. However, Dii did not resign, but selected the most promising approaches from the new findings. This also meant saying goodbye to unrealistic romanticism, false hopes and dreams – a painful path for many. In three to five years, the companies involved in Dii had been able to sound out whether the desert regions could be an interesting market for them and were aware that, contrary to what had been assumed in 2009, the current focus was on local supplies. The disillusionment was even a blessing, as the Dii team was able to work more realistically and practically. Due to the smaller number of partners, however, Dii had to live with less support from the corporate side. Only a few still saw the added value for their own business opportunities that Dii could offer as a universal pioneer in the MENA region. The expected fast and large business opportunities had failed to materialise.

In its first years, however, Dii had provided a wealth of valuable facts about the converging energy markets and the competitiveness and integration of renewables in the MENA countries and in Europe. It now had a unique geographic information system (GIS) and a database with detailed project-related information on all renewable projects in the region. It had built up a vast international network of interested companies, institutions, governments, parliamentarians, authorities, research institutes and NGOs. An unprecedented dialogue between players in different continents and countries had indeed been established. These long-term relationships between individuals and companies are probably the most important basis for sustainable business in the region. Professionals in MENA, Europe, China, Japan and the USA could now better assess the energy sources in the deserts, the needs of the local population and the art of doing business in Arab countries. Electricity from the desert regions was now of global interest, particularly in countries in sub-Saharan Africa. The long-standing lethargy in the Arab world was overcome, the renewables presentable and soon also competitive. 2014 was the last year in which PV and wind still needed massive government support. 2015 became the subsidy-free starting year for the expansion of renewables in the desert.

4.3 *The situation in the MENA region*

Where did the MENA region stand politically? Three years after the Arabellion one could not claim that the political and social situation had changed for the better. The political development in the MENA countries was very broad. Autocratic or rule-based structures still dominated. In Morocco, King Mohammed VI had survived the Arabellion well; in Egypt, Al-Sisi had taken power. With the exception of Yemen, the Gulf States remained relatively untouched by the outward upheavals. They were further steered by the monarchs and sheikhs with their families. In Algeria, the situation around President Bouteflika remained unchanged. Also in Turkey, Iran and Jordan the firm hand of the rulers asserted itself. In Tunisia, on the other hand, a vulnerable democracy unfolded. Libya, Syria and Yemen were involved in civil wars under very different circumstances in the Arab autumn. Iraq and Syria suffered under Islamic State (IS), which rapidly gained popularity.

A comprehensive analysis of the political situation would go beyond the scope of this book. However, it should have become clear that the energy issue in MENA can only be comprehended in the context of the complex interests of the individual Arab states, the regional forces, Israel, Turkey and Iran as well as the major powers USA, EU, Russia and China. The conflicts seemed to become more complex. It was no longer just about oil and gas, but also about migration, fear of terror, the IS movement, women's emancipation and rapid technological and media developments (Lüders 2015). In addition, xenophobia, especially against Muslims, was on the rise and populism was growing in Europe. An uncomfortable situation for Dii and its partners. Even though the rapidly growing Arab economies and energy markets were promising, due in part to rising population figures, business success was still hard work and could not be achieved without a deep understanding of the situation and strong local partners.

The energy market in MENA in 2014 was still based almost exclusively on fossil fuels such as oil, gas and hard coal. There were also plans in the various countries for large nuclear power plants, above all in Abu Dhabi, where 2,000 megawatts of nuclear energy were planned for 2010. The first power plant should be connected to the grid in 2019. The renewables were still only very rudimentarily represented: The hydropower plant at the Aswan Dam and more than 500 megawatts of wind energy in Egypt, some smaller hydropower plants in Morocco, wind plants in Morocco and Algeria and here and there CSP and PV plants.

In the dominant ruling circles, oil and gas were, so to speak, in the genes. Renewables were either hardly taken seriously or simply regarded as an expensive luxury. The population did not pay much attention to the topic. Energy was heavily subsidized and electricity was, apart from areas with technical bottlenecks, seemingly limitless. This explains why MENA is one of the slowest energy transition regions in the world. The realization that even in the MENA region one cannot rely infinitely on oil and gas became more and more audible in government circles in 2014. More and more projects for renewable energies were developed. The decisionmakers also recognized that energy cannot be subsidized indefinitely, which became increasingly clear, especially with the collapse of the oil price in the second half of 2014.

4.4 *Dii and its partners*

The collapse of the German CSP and PV industry after 2011 and the directly or indirectly related internal conflicts within Dii had a devastating impact on the team, the commitment of the companies and the Desertec initiative as a whole. It was announced in the German media that the mission announced in 2009 was a total failure. Desertec had 'failed and died', was a frequent read. It was not mentioned that the development of renewable energies in MENA was hopeful and that in the long term it was very likely that Europe would also benefit. Also, the public hardly learned that the Industry Initiative had grown into an international, even intercontinental organization. When it was founded, Dii GmbH had 12 shareholders with different interests and competences: The largest Algerian conglomerate Cevital, M+W Zander, which built production plants for semiconductor technology, Siemens and ABB, which had invested hundreds of millions in renewable energies, RWE and E.ON with their subsidiaries, the companies from the financial sector Deutsche Bank, Munich Re and HSH Nordbank as well as Schott Solar, Abgenoa Solar and MAN Solar Millennium, which focused on solar thermal power plants. It was a heterogeneous group seeking balance. This structure had changed in 2014. A number of companies had left the circle of shareholders for various reasons and new ones had been added, such as ACWA Power, Enel Green Power, First Solar, RED Electrica, Terna Energy SA, Unicredito and SGCC. Now it was grid operators, solar park and wind developers from the Mediterranean region and the Gulf States, as well as the State Grid Corporation of China, who shaped the initiative. The circle of partners had become more international and practice oriented.

The contacts with the Arab countries were as diverse as the countries themselves. Jamila Matar, head of energy affairs for the League of Arab States (LAS), had already made initial contacts at the announcement of Dii in Munich. Over the years, she had supported Dii from the bottom of her heart and brought it into contact with the key people in the region. In 2017 she joined the Advisory Board of Dii.

In the Gulf States, close contact was established with the Gulf Cooperation Council Interconnection Authority (GCCIA), the integral network operator in the region based in Dammam. Over the years, cooperation has intensified, culminating in a joint systems study in 2017 and 2018, also with Chinese partners, on the integration of photovoltaics in the region. It dealt with the question of how large quantities of PV electricity can be integrated into the GCCIA grid. An important condition for the absorption of PV electricity is the reinforcement and expansion of the transport networks. The Gulf States are already planning new connections to Egypt and Iraq. The same study examined a new international network connection, from the United Arab Emirates to India, mainly using a very long cable connection off the coast of Iran and Pakistan.

After contact with the Desertec Foundation was de facto interrupted in 2013, Dii worked closely with the NGO Germanwatch. Germanwatch was thoroughly concerned with the environmental compatibility of the planned CSP plants in Morocco. Local NGOs in MENA were still relatively rare. Talks about the interests of the population had therefore only been sporadic in the Maghreb countries and in Egypt. There have also been contacts with some parliamentarians in Mediterranean countries.

4.5 *The Rome Conference*

The group of 18 shareholders and 35 associated partners in the Dii circle had a service contract with the initiative until the end of 2014. Formally, Dii had been designed from the outset for a limited period of time to carry out the planned studies and to get the market moving. In the turbulent atmosphere from 2013 onwards, many people were less interested in remaining shareholders or partners of Dii after the completion of the studies. In 2014, the Dii team in Munich consisted of around 20 people with fixed-term contracts until the end of the year. The future of Dii was uncertain, in public it was already written off. That depressed the mood of the team. As managing director, Van Son had experienced a similar situation of general hopelessness and disorientation several times in the past. It was just then that it was important to offer new opportunities. The great confusion and disappointment also had a positive side. Ballast could be thrown out and the needs analysis for the continued existence of the industry initiative could begin. Only a core group was ready to continue the mission: Paddy Padmanathan (ACWA Power/ Saudi Arabia), Michael Geyer (Abengoa/Spain), Liangzhong Yao (SGCC CEPRI/China), Detlef Drake (RWE/Germany), Ernst Rauch (Munich Re, Germany), Jochen Kreusel (ABB/Germany-Switzerland), Christopher Burghardt (First Solar/USA) and Antonio Cammisecra (Enel – Res4Med/Italy) offered active participation in follow-up talks. In several sessions the following line crystallized:

- Dii possessed unique knowledge and a very valuable network in industry, the public sector and civil society and should, if possible, continue its work as a trailblazer for energy transition in MENA. The proposal was to change Dii from a temporary to a permanent organization.
- The first priority should be to accelerate renewables as pragmatically as possible in the MENA countries: *Local MENA Renewables First*. The answer to the question whether electricity or energy in whatever form would ever flow into Europe on balance should be left to the market.
- The main location of Dii should be relocated from Munich to the region or at least to the Mediterranean area.
- The leadership and composition of the team should reflect the needs of the region (local roots, locals in management).

This sounded logical and constructive to most of the participants, but questions regarding financing, location and organisation could not be solved. Some Dii shareholders therefore concluded that the company should be dissolved. After heated discussions among the shareholders, a formula for the continuation of Dii could only be found with great effort and skill of the chairman, Detlef Drake.

Media interest in the conference in Rome on 13 October 2014 was correspondingly high. A decision on the future of the initiative was to be made at the shareholders' meeting on the first evening. There was tension in the air on this warm Roman October night.

Van Son made three suggestions: First, dissolving Dii, second, slimming down Dii or as a last resort taking over Dii by the Desertec Foundation or Van Son himself. Around 11 p.m. ACWA-Power, RWE and State Grid of China/CEPRI decided to continue in Dubai in a leaner form. These three companies also agreed to increase the share capital. Although Van Son had already announced in September 2014 that he would move to a new position at RWE, the manager was asked to continue as CEO of Dii in 2015. At RWE, Van Son was appointed Director and Country Chair for MENA and Turkey to manage the local business from Dubai. The decision to continue Dii from Dubai was made quickly and with the full support of the shareholders.

Figure 23: Dii repositions itself after the shareholders' meeting. In the centre Thomas Altmann, CTO of ACWA POWER. Source: Thomas Isenburg

Thorsten Herdan, Head of Department of the Federal Ministry of Economics and Energy, spoke on behalf of the German Federal Government at the conference. The engineer alluded to the energy turnaround in Germany, but also called for a broader view. He confirmed the positive development of Dii's project and reminded the audience not to think too quickly, but to keep the planning horizon until 2050. There are strong signs that renewable energies can also be financially competitive, Herdan said in the course of his speech – only political tools need to be created. The demand for renewables is great within the EU. He mentioned the energy sponsorships with Tunisia, Algeria and Egypt. In these relationships we should try to learn from each other. It was the not right to give only instructions in one direction. Germany has founded Dii GmbH, now it will be continued together with Saudi Arabians and Chinese, the energy specialist summed up.

Germany wants to continue to participate, Van Son also affirmed. In his speech in Rome, he spoke of an establishment of the desert current in the countries of North Africa, where 70 projects now exist. Further steps would be the exchange on the continent and finally the intercontinental exchange in the EU-MENA region. The re-elected CEO compared the company history of Dii with the history of Romulus and Remus, who according to legend were brought up by wolves: "Before Dii existed, hardly anyone had any knowledge of Desertec. The great idea was neglected and ignored. Then the industry took care of the vision and she learned to walk." As Rome was not built in one day, so Desertec would not and will not be built in one day.

4.6 The move to Dubai

The plan was to establish a local office in Dubai in January 2015 with four members of the Munich team and to continue the work there. Shortly before the trip to Dubai, however, three of the four colleagues were poached by other companies. Only Ahmad Yousef, an expert in geographic information systems (GIS), landed in Dubai with Van Son. Another bitter setback. With the help of Thomas Altmann, Chief Technology Officer of ACWA Power, and friends and Innogy colleagues in Dubai, a small team was set up relatively quickly. In 2015 and 2016, however, activities remained modest.

Figure 24: New start in Dubai. Source: Dii

The focus was on the question which obstacles developers of solar and wind projects encounter in MENA and how these can be eliminated. Siemens came back on board as a partner. The enormous enthusiasm in the market for a new 200-megawatt photovoltaic, world record project of the DEWA (Dubai Electricity and Water Authority) and ACWA Power also moved other companies to join Dii. The United Arab Emirates, Egypt, Morocco and Jordan implemented attractive new projects. The first discussions about desalination on the basis of PV or CSP technology in the water-poor region arose. ACWA Power continued to establish itself as the driving force and soon became the market leader in the region. Masdar, which had already started building Masdar sustainable city and other projects in the CSP, wind and PV sectors in 2008, was also a high-profile player. In Germany, Gerhard Hofmann, Wolfgang von Geldern and Eicke Weber of the Fraunhofer Institute remained loyal supporters of Dii for many years. During this time, the acting CEO of RWE, Peter Terium, also supported Dii within the energy supplier from the Ruhr area.

4.7 *Dii takes action*

In 2016, Ismail Fahmy, a renewable energy expert, joined IRENA for one year as a local Dii manager. The agency is based in Abu Dhabi and deals with the implementation of renewable energies in various countries. Fahmy gave impetus to relations with local authorities and networked with new players in the private and public sectors with Dii. In MENA, they became better known. The annual *Dii Desert Energy Leadership Summits* and the guide *Who is Who in MENA and Turkey* (Dii 2017) published by Dii were very well received in Dubai. Numerous prominent representatives of the Gulf energy industry, such as Sheikh Ahmed Bin Saeed Al Maktoum, who acted as patron, attended the Leadership Summit. Saeed Al Tayer, CEO of DEWA (Dubai Electricity and Water), one of the most important managers in the region, and Peter Terium were the speakers. The participants came from Morocco, Egypt, the Gulf States, Turkey, Europe, China and the USA. The main topics were again: Why did MENA start the energy transition so late? What is the problem? How can the conditions for project developers of renewables be improved? The answers were simple. The first instance was to inform private and public stakeholders that renewables were already competitive where fossil fuels were not subsidised. Financing was not the problem, because capital was cheap in the market and a number of inaugurated investors accepted the risks. In relatively stable countries such as Morocco and the United Arab Emirates, the conditions for capital-intensive investments were able to keep pace with those in OECD countries. The internationally known and proven models PPP (Public-Private Partnership), IPP (Independent Power Producer) and long-term PPAs (Power Purchase Agreement) became established as contract constellations for renewable energies. In the beginning, government subsidies for wind and solar power were still needed. In countries with higher risks, capital costs and project development costs were usually higher. The international community provided assistance through promotional banks, grants or financial guarantees. Successful examples were created in Morocco, Tunisia, Egypt and Jordan. Germany again provided support at all levels, including financing by KfW and advisory and support services by GIZ, and indirectly by the EIB and the World Bank.

4.8 New momentum

In 2017 Cornelius Matthes, a former director of Dii, took over leadership of Ismail Fahmy's team. Matthes has long known the banking and energy world and had built up a broad network of industrial partners and other collaborations in the early days of Dii. Together with Uma Dagdelen, the former event and membership manager of MESIA (Middle East Solar Industry Association), he energetically revamped the industrial network. Fadi Maalouf, former Technology Director for the MW Zander Group (then a shareholder of Dii), joined the team as Chief Technology Officer.

Expert voice Fadi Maalouf (CTO Dii Desert Energy, Dubai)

Fadi Maalouf. Source: Privat

Growing up in Lebanon and working in Dubai, Fadi became exposed in 2017 to Dii's DP2050 'roadmap', which opened his eyes for what was believed to happen in the MENA countries. The study has given the starting signal for a turning point in the region toward a mega energy transition at regional and intercontinental level. DP2050 encouraged the energy community to rethink and reshape strategies for clean energy and interconnected power grids and Fadi became enthusiastic for Dii and offered to work on the most important prerequisites for success in the Arab world: workable, practical tools for project development that give people trust. Hence, he focused on the power of hands-on information, know-how, innovation ideas, and collaboration. A tight cooperation with Dii and experts resulted in a very popular *Dii Toolkit Initiative*. The initiative laid down a well-defined workflow to help developers make their projects more effective:

The Toolkit for renewable energy-based projects establishes a knowledge base enabling talent connection, opportunity creation, which enables implementation acceleration. Dii's Toolkit for 'Renewable Energy Grid Integration, Project Development and Industry Localization' is derived from Dii's partners as well as renewable energy industry expert professionals' know-how, experience, and best practices. The objective of Dii's Toolkit is to give stakeholders, both private and public, access to state-of-the-art measures and techniques (tools) which enable accelerated and smooth integration of large amounts of renewable energy into existing power grids and ensure tangible and durable benefits to the MENA region and beyond.

As of the first quarter of 2019, the toolkit initiative completed several publications covering a wide range of practical topics, such as: IPP PV Project Development Roadmap: '8-Phase Bankable Approach', Project Finance Management Plan for Utility Scale Solar PV, Business Plan Standard for IPP/EPC Renewable Energy Company, Pre-Feasibility PV Financial Model, and Pre-Feasibility Study Utility Scale Battery Energy Storage Systems (BESS) LCOS Financial Model and a so-called 'SunBurn Test'™©.

As 'proof of the pudding', a joint study was made with State Grid of China, Geidco and the grid operator of the interconnected grid in the Gulf Countries. The question was investigated indicatively how much PV power can be adopted in the power system and how the power system can capture synergies through a cable connection to India. The Gulf Cooperation Council Interconnection Authority (GCCIA) owns and operates the grid that interconnects the six GCC states. GCCIA is aiming to become a striving regional and global leader enabling energy trade and enabling a rapidly growing portion of renewables in the power mix.

The study was concluded in July 2018 and it took around ten months to finalize. The study addressed the benefits of implementing 1 GW solar PV plant (phase 1) which is considered a multilateral project of common interest (PCI). The analysed benefits included: environmental, social, economic, and energy security benefits. The study determined the technical, financial and legal approach for implementation. This included: site selection, concept design, Levelized Cost of Electricity and sensitivity, basic and detailed legal structure, tendering and procurement approach, project realization timeline, and project finance management plan. The study analysed the impact assessment of integrating renewables in the GCCIA grid and it used several schemes for detailed examination. The study also covered the export of renewable energy to nearby regional grids. A study case of interconnecting the GCCIA grid to the Indian grid was evaluated and the case included explorations of several interconnection schemes utilizing HVDC technology of up to about 800 kVDC and up to 8 GW capacity employing overhead lines and submarine cables.

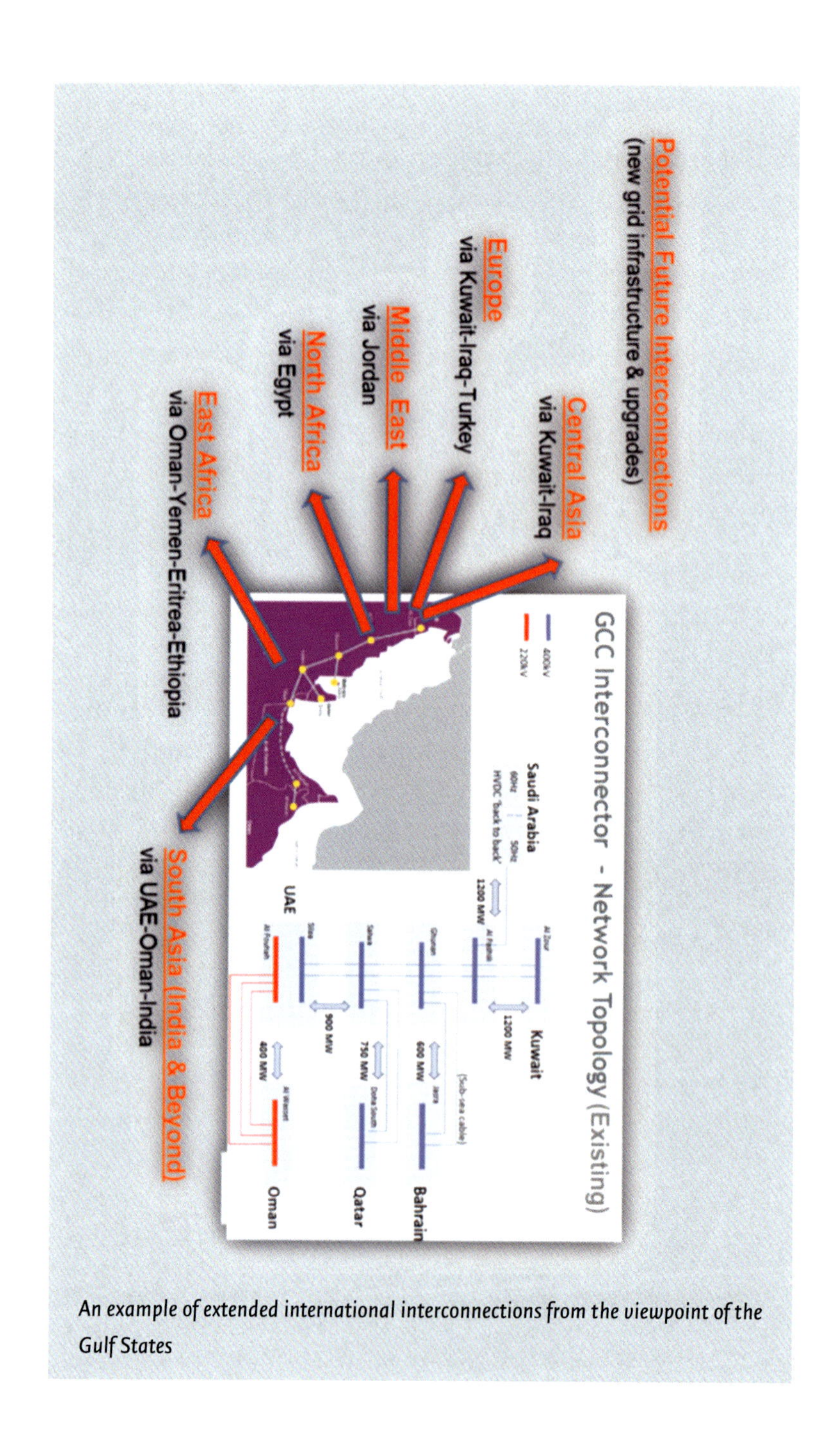

An example of extended international interconnections from the viewpoint of the Gulf States

The Advisory Board

Under the chairmanship of Frank Wouters, former Director of the EU-GCC Clean Energy Technology Network, former Deputy Secretary General of IRENA and before that Director of Masdar Power, a new Advisory Board with experts from the energy and banking sectors was set up in 2017. (Wouters 2018) Dii gained new momentum and flourished in the open and future-oriented culture of the Gulf. The circle of partners also grew again after many years. Through practice-oriented events and cooperations, Dii expanded its role as a pioneer and industry initiative in the region relatively unobtrusively. The work in the United Arab Emirates, Oman, Egypt, Saudi Arabia and Algeria deserves special mention here.

Figure 25: Cornelius Matthes, Frank Wouters, Liangzhong Yao and Paul van Son continue the work of Dii as supporting columns. The structures are becoming more international. Source: Dii

Expert Voice: Frank Wouters Public-Private Cooperation EU-GCC.

Frank Wouters. Source: privat

Europe and the Gulf have had a longstanding relationship and a ministerial-level dialogue was institutionalized in 1982. Energy, an important aspect of the relationship, is discussed every year in the Energy Experts' Group, comprising experts from the European Commission and the Secretariat General of the Gulf Cooperation Council (GCC). In their annual meeting of 2009, coinciding with the establishment of Dii, it was decided to establish a public-private sector network as a more continuous and practical instrument for engagement. The purpose of the EU-GCC Clean Energy Technology Network was to foster clean energy partnerships between the regions. The European Commission made budget available and issued a call for tenders, which was won by a European consortium led by the University of Athens. The GCC appointed a matching network, led by the Masdar Institute in Abu Dhabi. The platform came to life in 2010, dealing with renewable energy, energy efficiency, interconnection, CCS and clean fossil fuels. The Network engaged with academia, the private and public sector and fostered partnerships between the regions through study tours, events, trainings and conferences, publications and of course a web page. After the first funding period ended in 2013, the Network entered a new phase supported by a memorandum of understanding between the Masdar Institute and the University of Athens. However, the activity level dropped substantially due to the lack of funding, until the European Commission found some new money and revived the initiative through a second call for tenders in 2015. The Network was relaunched in 2016 and currently operates with a team of 3 key experts, Frank Wouters, Mustapha Taoumi and Ioanna Makarouni. The team of permanent experts is a new element in the Network, which has considerably increased the continuity, effectiveness and impact. Recognizing the market dynamics, the Network has also been given some more operational flexibility and is urged to join other initiatives for more leverage. A strong partnership has been established with Dii Desert Energy, which allow to join forces in the GCC-EU framework.

A good example of the rapid developments are two solar projects that both started around 2010, when Dii and the EU-GCC Network became active. One was a 100MW CSP project, the other a 100MW PV project. They were both initiated by Masdar, where Frank Wouters was in charge of the Clean Energy department, and they were both roughly equally expensive from the outset. The 100MW Shams I CSP project was commissioned in Madinet Zayed in 2013 after an international procurement exercise and yielded by some margin the most expensive electricity in Abu Dhabi at the time, even though the price was lower than comparable projects in Spain. The PV project faced some delays due to institutional inertia, but finally resulted in the 1.17 GW Sweihan project, with a power tariff of 2.42 $ct/kWh , making it then not only the largest but also the cheapest solar project on the planet. These kinds of tectonic shifts in the energy sector are hard to anticipate, but organizations such as Dii Desert Energy and the Clean Energy Network could provide open minded guidance, knowledge, scenarios and a platform for exchange of information for the benefit of private and public stakeholders.

Both Dii and the Network are advocates for international cooperation and one of the common future vectors are 'green molecules'. Low cost renewable electricity can be used to split water into its components, yielding green hydrogen. With world record low prices for solar power, ample available land and decades of oil and gas experience, the Gulf is predestined to become a major producer of 'green molecules', and more and more green energy will be traded across borders. Of course, electricity interconnection will continue to play a role, but there is growing interest in green molecules that can be used as feedstock, transport fuel and for industrial processes. The conversion of green electricity into hydrogen, ammonia or methanol furthermore enables bulk energy storage over seasons and very cost-effective shipping over long distances. Dii and the Network support stakeholders engaged in that promising new business in a joined-up approach.

Connecting people, connecting continents, connecting markets

That was the motto of Dii from the beginning, while its composition was still in flux. As already mentioned, Siemens once again joined the circle of supporters. The technology group was heavily involved in the wind, grids, grid management technology, desalination, cooling and digitalisation sectors relevant to Dii. Of the big supporters of the beginning, ABB, First Solar and the Max Planck Institute rejoined or had never left. Other former associate partners and shareholders were also considering re-entering the company. However, a new circle of companies that were active in the region is also forming. These include General Electric, Alfanar, Al Gihaz, Masdar, Belectric Gulf, Europagrid, Thyssen Krupp, Noor Solar Technology, Amana, Access Power and Krinne. Some smaller partners of the first hours have always remained loyal to Dii, such as Maurisolar, a small developer from Mauritania.

From the beginning in 2010 to date Cornelius Matthes, as Director Business Alliances in Munich and later as Senior Vice President of Dii Desert Energy in Dubai, was successful in making the partner companies of Dii feel like being part of a unique 'big family'. His deep knowledge of the industry communities in the region and his personal engagement have helped many (mainly small and medium size) developers and other companies to discover and capture business opportunities.

The expectations and requirements of the partners were also different in 2015. Manufacturers usually had a great interest in a growing market for their products. For project developers and investors, the question arose as to the special features of the market and obstacles to the construction of solar and wind power plants. The manufacturers and developers were joined by engineering firms and consulting firms such as Roland Berger. They were more interested in a long-term vision with concrete solutions for project developers and authorities.

The question as to whether it would be possible to earn money with the projects was a priority for investors and developers. MENA remained a difficult market. Not only foreign developers, but also local companies complained that it was difficult to find and implement good project opportunities. In the more stable, economically strong countries, there were several tenders under high competitive pressure. In addition, the contract volume per project increased rapidly. The list of qualified suppliers was thus limited to a few large ones with corresponding references and a strong balance sheet. The IPP structure basically anchors local companies and in many cases connects them with one or more international investors. In the politically, organisationally or economically more difficult countries there were many announcements, but fewer concrete implementations. The conditions were rarely clear and stable. Tenders were often withdrawn again, or tenders were evaluated as not comprehensible. The risks were great. This was reflected in risk premiums for investment decisions. Projects that have to achieve a calculated return on investment of 8 percent in a stable country should be able to demonstrate 15 percent or more in a high-risk country. As much as the dramatic reduction of the production costs for solar and wind energy had accelerated the development of the market, there was also repeated talk of a race to the bottom, in which developers sometimes took significant risks in order to deliver projects at sustainably profitable conditions. The EPC (engineering, procurement and construction) contractors, which deliver turnkey projects, also had to give up a significant part of their margins in this context. There was little risk buffer left in the event of problems. Similar to the situation with manufacturers a few years earlier, this situation triggered a significant consolidation of the market for project development and EPC.

Dii had tackled the complex task of paving the way for the biggest infrastructure project of our time. Their ideas and insights had a decisive influence on the discussion and also allowed their own perspective to become a new one. Over the years, the GmbH has stood out due to its enormous resistance to stress. It provides further answers to pressing questions and networks the energy industry across continents.

Excursion: Intercultural Cooperation

Not all potential suppliers know how to deal with local customs and business practices. At best one has prepared oneself for the local corporate culture, but the practical implementation is problematic for many European companies. Although there are great cultural differences between the Arab countries, they all attach great importance to supporting local interests first and foremost and letting business relationships grow slowly. This requires a great deal of composure and patience. More important than written agreements are personal relationships and the spoken word. Influential families often play a decisive role. Sometimes the military, or authorities in the field of energy utilities, then again advisors in the circle of decisionmakers who are difficult to assess. The range of behaviour is wide. It ranges from straightforward, as is often the case in Dubai, to intransparent and arbitrary. Everywhere, without local partners and good relationships with government leaders and ruling families, successful business is virtually impossible. It also helps to involve local investors and development banks, which can improve transparency in particular.

There are always examples of European companies that do not know or respect these unwritten basic rules. However, violations of the invisible code of conduct can have devastating consequences. On the other hand, one is rewarded if one understands the delicate interplay of give and take, patience and long-term friendship. European manufacturers such as Siemens, ABB and Leoni have been working successfully in the region for many years. Several smaller companies are also doing well in market niches. There are only a few German project developers, and even the major German utilities played hardly any role in the MENA region. However, the already mentioned local developers like ACWA Power, which is very well wired in the highest circles, or Access Power, a young, very dynamic company in Dubai. Successful international project developers in the region often come from Japan, France, Italy and Greece. The most active EPCs in the region have their origins in China, South Korea and Spain, and more recently India. The dominance of PV manufacturers is clear in China, although Firstsolar scores highly with thin-film PV technology. German companies are prominently represented in the field of wind turbines, converters and grid technology.

The second phase 'Desertec 2.0' finished

Dii had left behind the times of great public attention, but also the times of the doubters and moral crusaders like those in Munich at that time. There were no more exhausting struggles, but loose, inspiring and open discussions in all directions. The small team was aware that it would take time for the new Dii line to reach the minds of MENA's decisionmakers – and their understanding was what mattered. However, the new approach , 'no emissions, no nonsense, competitiveness of renewables will drive the market', was positively received by companies. Dii has thus completed its first task and is now positioning itself under the name Dii Desert Energy in Dubai and Munich as a sparring partner of the industry. One of the subjects under investigation is practical barriers in the market.

5. Prospects for emission-free desert energy: Desertec 3.0

About ten years have passed since Dii was founded. During this decade, much has been learned about the relevant energy markets in MENA and its environs. A differentiated look at the vision of desert power is necessary. The experienced observer recognizes a picture with light and shadow. The originally planned unique intercontinental *Energy Handshake* originates from German thinktanks and reflects a European approach. This meets the Arab structure and culture. In addition, technology and market realities for renewable energies are changing rapidly. You can be critical of Desertec's first ideas. The basic idea of developing desert energy was, is and remains valid. As a network of internationally operating companies, Dii has made desert power an issue in the media and among political and economic decisionmakers. A broad public in MENA and Europe was informed and sensitised. The processes have now become more concrete on the basis of rather mystical ideas. In the meantime, numerous successful solar power and wind projects have been carried out in the region, so that the development of these almost unlimited energy sources is no longer an illusion. Grid interconnections in the region and with Europe are a reality, and the way is open for desert electricity in 2018. The physical transport of electricity or the transport of gases or liquids produced with emission-free electricity over longer distances is controlled by the market and depends first and foremost on the cost and price differences as well as the demand in the various markets. There is no lack of capital, government programmes or economic feasibility. The perception of urgency is rapidly increasing in view of the increasing climate problems. The great art now is the completely new, much more attractive location in which MENA will be an 'emission-free powerhouse' for itself and for the world to enable and communicate widely.

The emerging voice of renewables

The international Renewable Energy Agency (IRENA) in Abu Dhabi plays an important role when it comes to global communication on renewable energies. It is an intergovernmental organisation that supports countries in their transition to a sustainable energy supply. To this end, the use of renewable energies is reviewed annually in the form of capacity studies and cost studies. The proposal for an international agency for renewable energy was made back in 1981 at the United Nations Conference on New and Renewable Energy Sources in Kenya's capital, Nairobi. IRENA was founded in 2009 with headquarters in the former federal capital Bonn and 70 Member States signed. It now has its headquarters in the eco-city of Masdar City in the United Arab Emirates. In the meantime, 153 countries and the European Union are members. IRENA announced that by the end of 2018 renewable energy generation capacity in Egypt was 2,351 gigawatts. This is about one third of the total electricity capacity installed worldwide. There are more and more convincing business cases for renewable energy, which the organisation cites as the reason for this. The largest share is accounted for by hydropower with an installed capacity of 1,172 gigawatts. This is about half of the total installed volume. Wind and solar energy, with capacities of 564 and 480 gigawatts respectively, accounted for the lion's share of the remainder in 2018. In 2018, the organisation published its fifth study on jobs in renewable energy for the fifth time. In 2017, it employed 10.3 million people worldwide. This was 5.3 percent more than in the previous year. Most jobs were in China, Brazil, the United States, India, Germany and Japan. These countries account for 70 percent of jobs. Most jobs are in the PV industry. Here, China has the lion's share. Employment opportunities here also increased in 2017. In the wind industry, there were fewer jobs in 2017 than in the previous year. (IRENA 2018). In publications IRENA has set out criteria for investments in renewable energies. In addition to the internationally binding results of climate protection negotiations, these are increasingly interesting business cases. These have been on the increase recently due to often rapid cost reductions for electricity from renewable sources. Depending on the region, these technologies can compete with conventional power sources. The globally weighted average costs for electricity from all renewable technologies, with the exception of CSP technologies, fall under the category of fossil fuels. For electricity production, these costs ranged from USD 0.047 to USD 0.167 per (kWh) in 2017. Onshore wind energy, where good resources are available, is now one of the most cost-effective sources of renewable electricity. Global weighted average electricity costs for onshore wind decreased by 23 percent to USD 0.06 between 2010 and 2017. At the cheapest auctions in Brazil, Canada, Germany, India, Mexico and Morocco, electricity costs for onshore wind power have fallen to 0.03 USD/kWh. The records for solar PV in Dubai, Mexico, Peru, Chile, Abu Dabi and Saudi Arabia were USD 0.03 per kWh.

The global finance enabler

The World Bank Group is also addressing this issue. Its original task was to help the states devastated by the Second World War. Now less developed member states are to be supported. The institution based in Washington sees good opportunities for trading in renewable energies in the Arab world. By 2030, around 39 percent of the energy mix could come from renewable sources. This would mean an installed capacity of 114 GW of solar and wind turbines. In 2017, the share of renewable energies was only 1 percent. Estimates suggest that a pan-Arab renewable energy market could be 280 to 290 GW in size. However, this will require enormous investments in the pan-Arab power grid. In addition to technical investments, it is important to develop appropriate rules for the market. This is why the World Bank Group and its partners have launched the Pan-Arab Regional Platform initiative. The aim is to promote trade in gas and electricity in the regions between the continents of Africa and Europe. The Arab electricity grids are to be connected to each other by 2038.

> *"Don't aim for a fixed plan for the next decades. Developing an emission-free market is an ongoing searching process with trial and error."*

There are no fixed timetables for the implementation of an emission-free energy supply. It remains a search process in every country with many possible paths. In addition, the global path following the Kyoto and Paris climate protection conferences needs well-functioning means of exerting pressure to reduce harmful emissions. The ETS (Emission Trading System) or CO_2 tax, which makes CO_2 emissions from fossil fuels a cost factor for coal and gas-fired power plants in various countries and regions, is not much more than a trend-setting start, driver of energy transition. Governments around the world are struggling with the introduction of these instruments, which are only accepted with much effort by the respective populations. In the MENA countries, no concrete, direct CO_2-reducing measures have (yet) been introduced. Governments and the market in MENA are still at the very beginning, but a breakthrough in desert power is still particularly visible in the low costs.

The review of the past ten years shows that the interested public has become considerably more competent in MENA and emission-free energy in the deserts. The lack of knowledge about cultures and economic conditions has not yet disappeared in all international players. However, news about the Arabellion, refugee flows, IS, upheavals, civil wars in the Arab autumn but also solar technology world records in Dubai and Abu Dhabi, on the Arabian Peninsula and in Morocco's Ouarzazate and reports about extremely low costs of wind energy in Egypt and Morocco have had their effect. In addition, there are new major projects in Egypt and Saudi Arabia with a volume of hundreds of billions of dollars.

MENA's deserts have supplied gas and oil for many years and could continue to do so for many years to come. This realization collides with the inevitable end of the fossil age. Desert power has long been state of the art and will play the leading role in energy supply in MENA in the coming decades. So, this also means that the oil and gas industry must find new prospects in good time...

The process of finding perspectives for the region is based on open discussions with local and international stakeholders. The main premise among the players experienced in this topic is that the MENA countries should first design their energy supply in an efficient, sufficiently open and low-emission manner and then become net exporters of emission-free electricity in a market economy manner. MENA will eventually become a solar and wind powerhouse.

From today's perspective, what are the prospects for the coming decades?

The four technical core processes of emission-free energyr 'power *generation/- conversion, storage, transmission/transportation* and *consumption*', are facing radical changes. In particular, the question is no lionger if 'volatile' energy sources sun and wind can take over from 'substantially less volatile' fossil and nuclear sources, but how, when and with economic benefits . In global practice, it has become increasingly clear that a balanced interaction of core processes makes a 100 percent share of renewables possible. Among others the global NGO, Energy Watch Group that show how regional and regional and global energy supply will become emission free. In particular sytems the leadership of Prof. Breyer of the LUT University from Finland provide the scientific rational for elekticity alone and will be enhanced in due course with `green molecule options'.

The prerequisite for urgently needed transformations is a profound rethink. The telecommunications and Internet industries have already experienced this process. In the future, energy supply will no longer be organised as a one-way street from producer to consumer. Similar to the Internet, the number of levels is increasing both bottom-up and top-down. There will be an interplay between decentralised as well as central and also autonomous, disconnected supply routes. Instead of long-term planning, the gradual replacement of fossil fuels by renewable sources will take place. Renewable energies are becoming the more cost-effective option due to rapid relative cost degression both in MENA and in Europe.

5.1 *Emission-free power generation*

Coal, oil and gas account for the lion's share of the MENA economies' energy mix in 2018. This cannot change completely in the short term. However, in the long run they have little chance of asserting themselves on the energy markets of these regions. They still often benefit from an intransparent market environment with government support. So far, climate costs have hardly been taken into account. But the greenhouse gas responsible for climate change will also lead to costs and liability risks in this region. The 'end of pipe' solutions for storing emissions are hardly practicable on a larger scale. Therefore, the storage of carbon dioxide with technologies such as CCS (Carbon Capture and Storage) or the incorporation of fossil-produced carbon dioxide into synthetic fuels such as methanol or kerosene will not clearly solve the climate problem. (Brown 2018)

The Paris Climate Agreement COP21 increasingly calls for the gradual substitution of fossil fuels by formulating the 1.5 degree target. As climate problems become more visible and urgent, one can expect stringent measures to follow. Finally, the planetary boundaries are crossed and extreme weather conditions become more visible. In the economy, especially in agriculture, the consequences of climate change are increasingly becoming a liability issue. Nuclear energy could theoretically offer a solution from the point of view of avoiding carbon dioxide emissions. On the other hand, there are serious disadvantages such as the high costs, the unsolved problem of final disposal and the long construction times of the power plants. In addition, there is a residual risk, as demonstrated by the reactor catastrophe at Fukushima in the high-tech country of Japan. However, nuclear power can produce electricity if consumers need it and cover the base load well.

However, the base load is hardly in demand during the day due to the greater supply of solar electricity. Nevertheless, several nuclear power plants are planned or under construction in MENA in 2018, including in the United Arab Emirates, Saudi Arabia and Egypt, the latter with Russian support. The very extensive and extremely cost-effective projects in the form of emission-free electricity from solar and wind power plants, which are already on the market or will soon be connected to the grid, reduce the opportunities for nuclear energy.

Approximately 20 years after the first thoughts on desert current, the nomenclature is changing and new technical processes are emerging. Electricity from renewable sources in the deserts is called green electrons. When green electricity is used to produce 'energy-carrying molecules' (e.g. hydrogen, methanol and methane) – a process known as *(green) power to X* and *power-to-gas* or *power-to-liquids* – we speak of green molecules (Olah et al. 2018). In the recent past, science, motivated by the challenges of energy, has been more intensively exploring this possibility. The electrolysis of water produces hydrogen, while hydrogen and nitrogen in turn produce ammonia, a starting product for fertilizers.

Hydrogen, a promising energy carrier

The hydrogen produced by renewable electricity has applications in various industrial sectors, such as energy supply, transport and the chemical industry. Hydrogen is an effective energy source. It can be stored, transported and converted into energy at will, but substances with complex chemical structures can be produced from the reactive gas. (Van Wijk 2017)

Hydrogen can, depending on its relative economic efficiency, partially or completely take over the task of fossil fuels and nuclear energy. This means that the existing energy supply infrastructure can largely be maintained even during the major transformation of the energy system. Fuel cells, which produce electricity when hydrogen and stowage material are converted into water, can form a central element. The hydrogen required is generated directly or indirectly by PV and wind power from renewable energy sources. Water electrolysis technology is still relatively expensive and therefore limited in its applications. However, the interest of global technology companies is high. (van Wijk 2018) The efficiently produced hydrogen can serve as a storage and transport medium for energy as well as a source medium for the chemical industry, among others. So far, the applications of hydrogen have been limited worldwide. The safe handling of hydrogen no longer poses a particular challenge. Economic conditions have also developed favourably and hydrogen is well on its way to assuming a prominent role in energy supply. Another option is the conversion of hydrogen into the simple alcohol methanol. For this purpose, it must be converted with carbon dioxide. Methanol can then replace fossil fuels in mobility. Heavy goods traffic and air traffic in particular are looking for such solutions.

Expert voice Badr Ikken, Directeur General, IRESEN (Institute de Recherche en Energie Solaire et Energies Nouvelles), Morocco

Badr Ikken. Source: privat

Morocco discovered relatively early that renewables would be needed to cover its energy needs and that export of power would become interesting as soon as prices levels in Spain respectively the European power markets would make its offerings competitive.

Presently, Morocco is leveraging its position on the top among other winners of renewables in the region, such as Egypt and UAE. As Morocco has no indigenous fossil energy sources it understands the need to hurry up the energy transition. Morocco aims to extend its first mover position also to Power to X processes, which may include three phases: a. The production of green electrons (from wind and solar), b. Production of hydrogen through electrolysis and c. Further chemical conversions.

If I take the example of the solar assets in Ouarzazarte of 570 MW, it came at a cost of around 2,5 Bln Euro. Conversion into Hydrogen and then Ammonia (NH3) would require another investment of about one Billion Euro. Power to X would open up huge opportunities because the value of power fuels is 4 times higher than electrical power alone! The costs of one tonne of Ammonia produced with PV in Morocco would be nearby 350 € (based on 4€ Ct/KWh in 2019). The cost level of wind parks in Morocco is presently even lower (about 3€ Ct/KWh). Cost of electricity amount to about 65% of the total Ammonia production costs and the water need for producing Hydrogen is rather negligible (1 liter for 1 tonne of Hydrogen).

Morocco has the largest Phosphate reserves in the world and the OCP Group is the largest global phosphate exporter. NH3 is among other the basis commodity for producing fertilizers, which up to date hast to be imported. The classical conversion method to produce one tonne of Ammonia is halfway between 200 and 250 €.
OCP could deliver Carbon monoxide to produce Methanol from green Hydrogen. Green Methanol is a globally tradable power fuel, which could be produced competitively soon. It would be highly interesting for transport sector, aviation and space flights.
In the frame of the German Moroccan Energy partnership, together with three Fraunhofer Institutes, we have conducted feasibility studies on this subject. They should lead to a Road Map until 2030. Morocco established also a national ‚Power to X' Commission for this purpose.
We are also building up a national research platform for green hydrogen and green ammonia pilot projects to set the path for a power to X industry, based on the same model of the solar research platform, the Green Energy Park in

Benguerir, which was recently inaugurated by His Majesty the King Mohammed VI, with electrolysers from 1 to 10 MW. The cooperation with Fraunhofer Gesellschaft will lead to their first Fraunhofer Institute in Africa jointly with OCP and the Polytechnic University Mohammed VI (UM6P). The Fraunhofer Mazagan Lab will support the industrial upscaling.

OCP, UM6P, IRESEN and Fraunhofer have already joint forces, and we are looking forward for more north-South cooperation, involving industrial partners like Siemens, Thyssen Krupp, which have a long history in the field of electrolysis and Haber-Bosch processes to produce hydrogen and ammonia.

Provided our energy supply will be near emission free, Morocco will manage much larger Power to X processes and, hence, keep its number one position in the region.

Finally, we are looking in Morocco at the demand side, that includes in the first place the fertilizers industry and next step would be the energy sector in Morocco. Later on the export to Europe would be feasible via existing gas infrastructure and maritime transport. This will positively impact the regional geostrategy and will contribute to the creation of a new win-win partnerships: European clean energy union.

The study was concluded in July 2018 and it took around ten months to finalize. The study addressed the benefits of implementing 1 GW solar PV plant (phase 1) which is considered a multilateral project of common interest (PCI). The analysed benefits included: environmental, social, economic, and energy security benefits. The study determined the technical, financial and legal approach for implementation. This included: site selection, concept design, Levelized Cost of Electricity and sensitivity, basic and detailed legal structure, tendering and procurement approach, project realization timeline, and project finance management plan. The study analysed the impact assessment of integrating renewables in the GCCIA grid and it used several schemes for detailed examination. The study also covered the export of renewable energy to nearby regional grids. A study case of interconnecting the GCCIA grid to the Indian grid was evaluated and the case included explorations of several interconnection schemes utilizing HVDC technology of up to about 800

At first, Dii concentrated on large facilities. However, the modern energy supply begins in the homes of the people. PV-based power generation capacities will be built on windows, walls and carports. In the hours of the sun they produce green electrons. The consumer will cover part or all of his own demand for electrical energy through them. It can deliver surpluses to the network. Through networking, the supply and demand of households can be balanced. In the course of time, several hundred million electricity suppliers will emerge in MENA. Cities, agricultural areas, lakes and industrial plants produce green electrons and molecules. A new dynamic is created. The generation of electricity in the future will ultimately be determined (almost) entirely by the supply of renewable energy. The market will adapt flexibly to this offer, with consumption at the forefront. The road to go will, thus, require a total rethinking from the traditional 'Fossil based satisfaction of the consumers' to 'Smart adaptation of all market players to forces of nature'.

5.2 *Energy storage and smoothing of consumption curves*

Growth in renewable energy has long been held back by high costs and volatility due to dependence on wind and solar radiation. Efficient energy storage systems to compensate for fluctuations in renewable electricity production are not yet widely available and are not needed in the fossil-based energy industry. This situation will change due to a growing share of renewable energies in the energy mix. In the desert regions there is a large energy demand for cooling, analogous to heating systems in colder regions. Ice produced cheaply with PV electricity can take over this task. The cold water or ice reservoirs smooth the demand for electrical energy and push it into the sunshine hours. This can contribute to cost reduction. The cost advantages are reinforced by lower prices for PV electricity. In cities and settlements of the MENA region, systems for district cooling, block cooling or individual cooling are used. Further developments will be visible in batteries, hydrogen storage and other methods of energy storage. Electric cars are still rather rare in MENA, but first experiences in the United Arab Emirates are promising, and in Morocco there is even a rally for electric cars. The batteries of electric cars become an integral part of the energy supply through controlled charging and discharging processes. The competitiveness of the (new) energy storage will be determined by the storage costs per kilowatt-hour for defined storage cycles compared to the price or tariff fluctuations from supply and demand of the electricity market.

5.3 *Energy transport*

Some MENA countries rely on imports of energy, others on exports. Electricity transports across national borders are still very modest. Fossil carriers of primary energy are expected to be replaced over time by green molecules such as hydrogen or biomethanol and synthetic hydrocarbons produced from renewable electricity. Then an adaptation to the existing infrastructure is necessary. Based on the technology of LNG tankers, which are designed for the transport of liquefied natural gas, hydrogen can be shipped worldwide. All MENA countries can thus become exporters of green molecules; an elegant approach. The transport and exchange of electricity has so far been technically and politically limited. An extension of the power grids in densely populated areas is very expensive and hardly possible. However, electricity transports in the form of direct or alternating current have become cheaper in recent years. The network expansions in MENA that are necessary for long-distance transport are responsible for about 15 percent of the energy costs if they can be built without excessive social counterforces. The transport of green molecules could be cheaper because their energy density is higher. Existing gas infrastructure could also be used. This is why power connections over long distances, including undersea lines, are not the best solution everywhere. The transport of electricity will increasingly have to compete with the transport of green molecules. China in particular is relying heavily on concepts for long-distance power transmission. Dii Desert Energy is integrated into the Chinese dominated organisation GEIDCO (Global Energy Interconnection Development Cooperation Organisation). This organisation is committed to intercontinental electricity connections in all regions of the world. The GEIDCO plans inspire network operators around the world. In the long term, electricity transport extensions in the Mediterranean area will be a prominent part of the plans for network operators in Europe (ENTSO-E) and the MENA region (MEDTSO/GCCIA). Interconnected grids become the platform for the converging emission-free electricity markets in MENA and Europe. Dii is conducting studies with GEIDCO and local grid operators on the possibility of integrating large PV and wind capacities and on new grid connections in the form of power cables between the Gulf States and India.

5.4 Flexible consumption and consumer choice

Rate of consumption is a key component in the energy supply of the future. The phenomenon of load control has been known for a long time. In MENA, load control has so far been of little relevance, apart from acute supply bottlenecks, because of the energy supply provided by easily controllable oil and gas power plants. That will change dramatically. The natural fluctuation of energy sources requires not only storage and transport systems, but also a sophisticated proactive adjustment of consumption to the permanently changing supply. Load management will become more decentralised and will be stimulated by price signals from supply and demand. Countries such as Denmark have had good experience with dynamic electricity and gas prices for some time. Such experiences can be very useful for the MENA region. Particularly in MENA, the flexibilization of tariffs towards energy prices that are continuously adjusted to supply and demand will offer very great advantages. Only in this way will it be possible to control a smooth adjustment of all types of consumption to surpluses or bottlenecks in the market. The economically attractive adjustment of consumption is particularly possible through the optimization of energy-intensive industrial processes. The adjustment of consumption by large consumers in periods of high electricity prices leads to a reduction in costs. Examples of this can be found in the metal industry and seawater desalination. Major consumers can use the optimisation potential offered by corporate sourcing. In addition, a local combination of PV and solar thermal can further improve the flexibility and service life of a system. It also makes sense to involve consumers more actively in energy transition. They should be able to choose green electricity. Since the physical product cannot be distinguished between green and non-green, it can be made transparent by green certificates, so-called IRECs. Certificates thus give electricity customers the option of promoting renewable energy plants.

5.5 *Developments in the MENA countries*

Today, it is impossible to predict how the countries of the MENA region will develop politically, economically and socially. So far, the conditions and processes in the individual countries have been different. Politically, MENA knows the whole spectrum from great stability accompanied by prosperity to civil war and disintegration. The oil and gas-rich region is the scene of geostrategic conflicts between major powers and trade interests. The different legal frameworks make cross-border trade difficult. This also affects international developers and investors. For each country, even inland, they need elaborate legal opinions and specialized legal knowledge. The synergy potentials are used little. A turnaround in the form of strong pan-Arab cooperation or a politically profiled community of Arab states is not in sight. However, there are clusters, such as GCC (KSA, Kuwait, Bahrain, Qatar, UAE, Oman) and Maghreb (Morocco, Algeria, Tunisia), where there is coordination between member states on energy, despite differences

Expert voice Mohamed Al Ramahi CEO of masdar

Mohamed Al Ramahi. Source Masdar

Established in 2006, Masdar, or Abu Dhabi Future Energy Compamy, is a global leder in renewable energy and sustainabl urban development. Masdar means source in Arabiv. Our mandate is to help maintain the leadership of the United Arab Emirates (UAE) in the global energy sector, while supporting the diversification of both its economy and energy sources for the benefit of future generations.

While sustainability challenges are global, developing countries often feel the damaging effects of climate change most acutely. The prosperity that has come from abundant oil and natural gas resources has helped the United Arab Emirates to mitigate the human impact of its arid climate. However, the importance of living in balance with the natural environment has been a guiding principle since its establishment.

Sheikh Zayed bin Sultan Al Nahyan, the founding father of the UAE, was a passionate conservationist. It is his legacy in sustainable development that we carry on today at the Abu Dhabi Future Energy Company, also known as Masdar, the Arabic word for "source".

Hydrocarbons are quite literally the bedrock of our national wealth in the UAE, but they are finite. Knowledge and human capital will be the most sought-after currency in the future.

As if to underline the importance of economic diversification in the national development strategy, His Highness Sheikh Mohamed bin Zayed Al Nahyan, Crown Prince of Abu Dhabi, famously remarked that the UAE will celebrate the day it produces its last barrel of oil.

It therefore comes as no surprise that the UAE was the first country in the Arab world to set a renewable energy target and its first signatory of the Paris Agreement. With social, economic and environmental imperatives underpinning the energy transition, Abu Dhabi took the bold decision to position itself at the forefront of the global effort to enable a more sustainable world when it launched Masdar in 2006.

Over the last 13 years, Masdar has combined its unique access to talent, capital and renewable energy resources to create a new source of economic value for the UAE over the long term.

Our business model is straightforward. On the one hand, we are helping to commercialise the latest technology in solar power, wind energy, sustainable real estate and waste-to-energy through its application at scale here in the

MENA region and around the world. On the other, as part of the Abu Dhabi sovereign wealth fund Mubadala Investment Company, we are striving to deliver on the sustainability targets of the UAE, and indeed the sustainability goals of every country in which we work, from the United Kingdom to the nations of the Caribbean.

In 2008, Sheikh Mohammed Bin Zayed conducted the formal groundbreaking of the world's first low-carbon city. Masdar City in Abu Dhabi is today the permanent home of Masdar, hundreds of other companies, both multinationals and local start-ups, and thousands of office-workers and residents.

It is also the location of the International Renewable Energy Agency, the first international government organisation based in the Middle East, and the Masdar Institute, a home-grown research institute specialising in clean technology and future energy solutions.

A template for sustainable urban development, where each new building achieves higher levels of power, water and waste efficiency than the last, Masdar City is undoubtedly one of our most iconic projects.

It is showing that sustainable real estate works, and is affordable, in a region where the built environment consumes as much as 80 per cent of the electricity generated. Not only that, the technologies and best practices employed in the construction of new buildings are helping to retrofit existing and less energy-efficient buildings elsewhere.

Without question, the bold sustainability ambitions of the UAE, which aspires to produce half of its domestic power needs from clean energy sources by 2050, are encouraging other countries in the MENA region to invest, as illustrated by the success of Abu Dhabi Sustainability Week, today one of the world's largest sustainability gatherings.

Hosted by Masdar each January, the event receives around 38,000 attendees from 170 countries and this year announced business deals worth an estimated US$11 billion. It also hosts the annual award ceremony of the Zayed Sustainability Prize, which recognises outstanding achievement in the five categories of health , food, water, energy, and high school education.

The Kingdom of Saudi Arabia is becoming a fast mover of 'Desertec 3.0'!

Expertvoice Peter Terium, NEOM Energy, KSA

Peter Teroiurn. Source: Neom

NEOM is a perfect example of giving shape to 'Desertec3.0' by building the energy world of the future in a cluster of new cities. First of all it is about building a sustainable living environment in the North-West of Saudi Arabia. In that area the largest carbon free energy system that has ever been built in the world is at the initial stage of being rolled out. That major development is well feasible due to solar irradiation conditions that are among the best worldwide and due to above average wind conditions close to the Red Sea. Emission free energy production costs at NEOM will, thus, be at very competitive costs, not only for the local and regional market, but eventually also worldwide. In order to make that work we are attracting 'brains', talents and hands-on people to work with us together here at NEOM.

Due to the protected special status of NEOM within the Kingdom of Saudi Arabia its legal system, regulations and market design will explicitly be shaped for a (net) emission free market. That will be completely new in the MENA region and it will be unique in the world. Saudi Arabia has not only a huge solar and wind potential, but it has also the potential to build up relevant, competitive manufacturing industry for solar panels, thermal energy technology, wind turbines, electrolysis, storage, transportation infrastructure etc. We believe that the future energy needs will not only be satisfied by 'green electrons', but also through 'green molecules' produced from solar and wind power. Green molecules in whatever molecular form will, thus, conquer the energy markets worldwide, accelerating de-carbonization. Green hydrogen has presently our priority, for which we will be starting a few pilots. We are, thus, dedicating much attention to all aspects of the value chain, HVAC and HVDC power grids including interconnections with neighhours, green hydrogen production, different kind of storage applications, green gases compression/fluidizing, green gases grids/shipping, smart (micro) grids for the cities and so on. All of this has its place in the energy masterplan at NEOM. I feel honored to be part of this historic and brave turn in a traditionally fossil dominated region for the next generations to come.

Although there are more and more renewable plants, decarbonisation in MENA as a whole is still progressing slowly. In 2018, only a few percent of the energy supply will be emission-free. After a whole series of flagship projects have proven their practical suitability under desert conditions, further projects are examined in detail in business cases. Feasible projects are emerging or can be expected in areas such as solar, wind, grid expansion, desalination, storage, electrolysis and electric transport. At the system level, digitization, Internet of Things, smart grids, cybersecurity and energy trading will ensure the integration of renewables and the optimization of the energy chain. For these tasks there are many committed young people in countries like Morocco and Tunisia. The arguments in favour of further subsidies for fossil fuels, nuclear energy and, ultimately, renewables no longer apply. In all projects and investments, the benefit for the local population is an important condition. Governments are rightly pushing for jobs in the manufacture, maintenance and operation of equipment. New industries can establish themselves in the surrounding area. This will push back investment in fossil technologies: Oil or gas producing countries must prepare for a shift from jobs in the oil and gas industry to renewables, as the example of Saudi Arabia shows.

Subsidies in the area of fossil fuels, including through electricity prices, have already been gradually reduced in all MENA countries. This process is difficult because it affects the populations financially. There is no alternative to subsidy dismantling. Competitive renewables no longer need subsidies. A development that is clearly visible in the tenders for PV systems in the United Arab Emirates and offshore wind in Northern Europe. However, a promising technology can be supported by the state for a limited period of time. Renewables will find their way into the market as a result of increasing competitiveness. Added to this may be indirect support through emissions trading. The expansion targets for solar and wind energy in MENA and Turkey are approximately 60 gigawatts by 2022. They do not necessarily have to be implemented but are important orientation points for market participants. Countries are rewarded for their efforts with concrete strategies for implementing renewable energies if they also monitor progress. In addition to IRENA, international agencies such as the International Energy Agency (IEA) can transport such a process to the global public. In addition, the Arabic League, the Regional Center for Renewable Energy and Energy Efficiency (RCREEE), the Union for the Mediterranean (UfM), the Mediterranean Transmission System Operators (MED-TSO) as well as the Gulf Cooperation Council Interconnection Authority (GCCIA) and other institutions in the Mediterranean and Gulf States can support and strengthen the purposeful target management of governments.

The technical conditions have improved considerably since the foundation of Dii and difficulties have been avoided. In the meantime, PV systems and wind turbines can be installed in the entire MENA region without major technical problems. The rotors of wind turbines are already being produced in Morocco, and soon in Egypt as well. In several MENA countries, CSP components or entire PV panels are created on the basis of imported solar cells. In contrast to coal-fired or nuclear power plants, the construction times for wind and PV projects are only a few years from planning to implementation and can be extended in each case. A critical aspect of solar thermal energy is water consumption in the deserts. Air cooling has now become established, although outside temperatures of up to 50 degrees Celsius greatly reduce efficiency.

An important criterion for the chances of realising a renewable energy project in North Africa is its financial viability. Investors demand a clear business case with a calculable return on investment. In most cases, projects are developed with one or more partners with project funding. Commercial banks and development banks require precisely comprehensible conditions for the granting of one or more project loans. They insist on bankability. Donors insist on a long list of evaluation aspects for project funding. Country risks are particularly important in the analysis of obstacles. Some risks are clearly quantifiable, others are estimated by investors. In the recent past, the MENA countries in particular have attracted attention through negative news and crises. This has deterred investors, because they demand a climate of stability for their commitment. Countries can benefit from the support of the EU and its Member States, as well as China and Japan. International institutions can also help to reduce the risk factors for financing. The first successful projects have led to regulations and standardisations in the tenders. The Independent Power Producer (IPP) structure enables several partners to participate in the production of renewable electricity. Power purchase agreements with the buyer of electricity lead to grid connections, financing and training. In addition, a circle of local and international consultants, engineering companies, developers and investors has formed in MENA who are familiar with the conditions and professionally contribute to the feasibility of projects.

As already described, hydrogen can become a big winner in the radical transformation of the energy system. Hydrogen is the most common element in the universe. Today we know that hydrogen forms more compounds than any other element. It is therefore a promising starting material for a wide variety of processes. Concepts for a hydrogen economy in which the energy can be transported or stored in the form of liquid hydrogen have already existed for several decades. Especially at the beginning of the 1970s a lot was published about it. Examples in the form of underground tanks and pipelines in the USA, South Africa and Europe are already known. As an automotive fuel, hydrogen has the advantage that more energy is released per unit weight of fuel than with fossil fuels, and no carbon dioxide is emitted. It can also be used either by fuel cells or internal combustion engines. An addition to the city gas is also possible. Recently, hydrogen containers have become even more reliable. With the help of renewable electricity, hydrogen can be produced by electrolysis. Like the conversion of gas into electricity and water, both processes are expensive because they are based on membrane technologies. By 2018, they can already be easily integrated into energy systems. As the demand for hydrogen increases, more electrolysis units and fuel cells will be needed. As in the early days of PV systems, this can lead to mass production accompanied by a reduction in production costs. Although closer system studies would be helpful, Dii expects hydrogen to become competitive in the energy market. This technology could then be able to substitute the primary energy sources coal, oil and gas in the form of solar fuel.

It will then assume the role of storage and transport medium for energy converted from renewable sources. For this purpose, the gas can be produced at decentralised locations and supplies electricity when it is needed. The trend towards decentralised generation cannot be stopped. Many consumers become producers, so-called prosumers. In the future, the green electrons will become extremely cheap during the day due to the simultaneous supply of many millions of PV and wind turbines. Markets should become more transparent and open, but far-reaching liberalisation is not a must. The vertically organized DEWA is a very successful example of innovation organized by a central management. Market liberalization and price transparency are not yet very important in MENA. One of the key components of the radical energy revolution is sector coupling: green electricity is used to reduce the use of fossil fuels in other sectors, such as chemicals, energy supply, transport and real estate. This opens up further opportunities for countries that are in a position to produce renewable electricity in the desert. If they meet the organizational requirements, they can become the winners of the future. How this works in the region is demonstrated by desert cities like Dubai, which are able to intelligently manage the four process elements of the energy system transformation. Climate-friendly and energy-rich oases are created here. This has a positive effect on the living conditions in the cities. Whereas in the past low energy costs were linked to oil and gas sources, in the desert country this function is now taken over by renewable sources with significantly price-reduced technology. The green molecules and electrons can be exported, traded and exchanged internationally, intercontinentally and sometimes even globally, analogous to the trade routes of fossil raw materials. In the past, the MENA region has assumed the role of a solid trading partner for fossil fuels. Now it can become an emission-free energy supplier. For this purpose, the Dii study DP2050 had already identified potential sources of income in the billions. The extension of the spectrum by *Power to X* techniques could strengthen this prognosis of the Dii study. The initial Desertec idea was still very fixated on power transmission. In the meantime, the cognitive processes have progressed further. Electricity transmission is expected to remain an option for connecting energy markets. However, the expensive HVDC cables are only worthwhile if a concrete business case exists. Also, not everywhere the construction of new power transmission lines is possible and economical. Alternatively, the green molecules can step in by transporting energy by ship or gas pipeline. This broadens the perspective for electricity exports, based freely on the original considerations of Desertec. This in turn provides options for high-quality value-added processes in MENA, promotes local industries and creates urgently needed jobs for the many young people. However, such developments have their price. The framework conditions for this are political and economic stability, good organisation, training and education for the people on the local labour markets. As Europe is under great immigration pressure from the region, it should have an interest in improving the quality of life in MENA. The economies of the MENA region can become emission-free in a few decades' time and, if developed accordingly, attract energy-intensive industries attracted by very low energy and labour costs, and also become net energy exporters. These are favourable location factors. Interested companies can therefore either relocate part of their factories to the desert regions of the world or import green molecules from the MENA countries. The topic will occupy the strategy departments. The decisionmakers of the industrial groups will not be able to avoid the price reality of emission-free energy. In addition, in contrast to the desert, the space available in Europe for the expansion of renewable energies is limited. In 2018, for example, there is already noticeable resistance to the construction of further wind turbines in countries with advanced wind energy expansion.

In the MENA region, renewable energies can contribute to reducing tension. Unlike oil and gas wells, they are evenly distributed throughout the region. This reduces the probability of conflicts over the power factor fossil energy sources. Renewable energy sources are freely available and are considered to have potential for democratisation. They are also a basis for water produced with renewable electricity. The adequate supply of water is a big issue, because in dry areas water means power. In this respect, too, renewable energies will have a positive, stabilising effect on the region. The exchange of energies between countries and continents leads to synergies, as system studies show. This should have a positive effect on world trade. Renewable energies connect continents, countries, governments, companies and people.

So far, economic growth has been based directly or indirectly on the import or export of fossil raw materials. This situation will change dramatically due to the great transformation of the energy system. The Arab Emirates, with Dubai, Abu Dhabi, and Morocco at the forefront, recognized the signs of the times early on. They lead the way to an energy supply based on indigenous renewable sources. The export countries Algeria, Saudi Arabia or even Qatar have the luxury that the income from petrodollars keeps flowing.

In times of low oil prices, however, it is no longer as simple as it used to be. Saudi Arabia is also feeling the pressure of change. It is clear that the fossil age must come to an end. They can use their excellent financial basis to turn energy systems around. For both energy importing countries and energy exporting nations there is no alternative to this energy turnaround. Local limited energy consumption can well be met from renewable sources such as solar, wind and various other renewable energy sources. This allows buildings and smaller businesses to become energy self-sufficient. In countries such as Morocco and Jordan, there are programmes to provide mosques with electricity with the help of solar cells and renewables. This is a strong impulse for the Muslim population. The exchange of large amounts of energy requires a suitable infrastructure. The task can be performed by electricity grids, gas grids, ships or alternative means of transport. Therefore, a completely new competition is to be expected. Regardless of the energy source selected – hydrogen, electricity, synthetic gases or liquids – the bottom line is the price per megawatt hour of energy delivered to wholesalers or end consumers.

An important aspect is data management, because the interaction of generation, storage, transport and consumption creates a gigantic wealth of information. This is required for the secure process control of the market participants and the infrastructure driving authorities. This will overtax the known methods of information management and process control. Therefore, techniques like Dynamic Pairing, Big Data Management, Blockchain, iCloud, Green Coin and other modern methods will find their way into the energy world. The first examples are already being tried out in Dubai.

Whether – for rational reasons – investments will still be made in fossil plants in the coming decades depends on whether the Risk-Adjusted Return On Capital (RAROC) on such plants remains interesting for even longer. So far, the energy industry has been pampered worldwide with lavish returns. Such returns are no longer realistic. The new solar and wind turbines in MENA still offer only very modest returns. Developers can be happy when 6-8 percent return on investment is achieved. Specific country risks apply to investors and developers. They can be reduced through many years of contact with local decisionmakers and with political support through targeted state-supported risk protection.

The Saudi company ACWA Power is one of the first project developers in MENA to benefit from the developments. It is a company with a strong vision, with courage, supported by shareholders with deep pockets and good connections to the local rulers and development banks. ACWA is able to open doors, take greater risks and negotiate strategically. This is the only way to reduce project risks. Each country has its own methods to support the exporting companies of the country. For example, German companies can strengthen their position through Hermes guarantees, payment arrangements and other mechanisms, but German policy is rather cautious here, as are German plant builders and investors. Chinese, Japanese, Korean or French companies are finding it easier to do this with the strong political support of their mother countries.

The MENA market for renewables has been slow to get off the ground. Due to the many professional tenders, however, it is increasingly possible to achieve the standard of the leading industrial nations. It seems possible that renewable energies will push fossil primary energy sources out of the market faster in MENA than in other regions. But only if political stability is achieved in the long term – because the political risk factor weighs considerably heavier than the market and economic obstacles.

Expert voice: Prof. Dr. Ad van Wijk (TU Delft, Netherlands)

Ad van Wijk. Source: privat

The deserts will undoubtedly supply the world with energy

The MENA region today is supplying the world and especially Europe with their abundant resources of cheap fossil fuels, oil and gas. Saudi Arabia, United Arab Emirates, Kuwait, Qatar, but also Libya and Algeria and others still have vast reserves of oil and gas. Oil and gas are transported through pipelines or by ship around the world. More and more, gas is transported by ship as LNG (Liquefied Natural Gas), but especially Algeria also transports gas by pipeline to Europe.

The transition from oil and gas to renewables seems obvious if one considers the vast and endless solar and wind energy resources in this region. Less than 10% of the surface of the Sahara Desert covered with solar panels alone could produce 155,000 TWh or 556 EJ. This was the world's total primary energy consumption in 2016. Wind energy could be mobilized on top of that.

So, this region can easily supply the world, also in the future, with cheap and reliable renewable energy, whereby prices may drop below 1 dollarcent per kWh for solar and perhaps also for wind in the foreseeable future. The challenge is, thus, how to bring this cheap renewable energy to local and remote consumers.

Hydrogen as the circular energy carrier for large scale transport and storage

The original Desertec idea was to transport energy from the deserts, by power infrastructures to the demand in the region and to Europe. However, recently hydrogen has emerged as a highly promising carbon free energy carrier. Hydrogen can be produced via water electrolysis from renewable electricity. It can be made liquid, converted to ammonia or bound to another element to transport it by ship over the world. And storage of hydrogen in large quantities is possible in salt caverns and maybe in some empty gas fields too. Essentially it is a circular energy carrier and together with electricity it will constitute the future carbon free energy carriers in a fully renewable energy system.

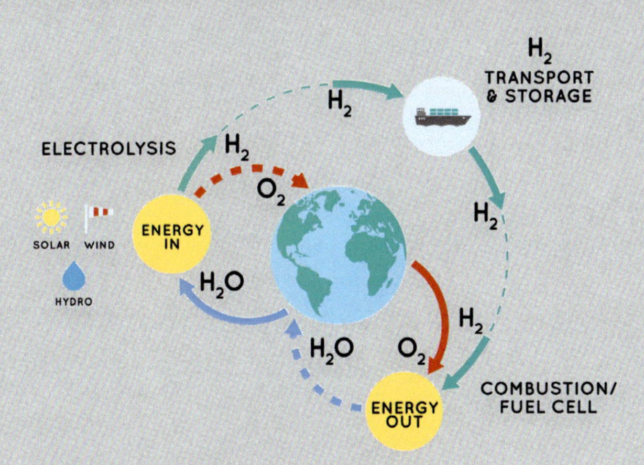

Hydrogen as the circular and carbon free energy carrier for the future

Hydrogen Pipelines for transport of carbon free energy e.g. from MENA to Europe

Today, Europe imports natural gas mainly from Algeria, with several pipeline connections to Spain and Italy. A transport capacity of about 60 GW via pipelines is already in place. For both the MENA region and Europe it would be very interesting to jointly unlock the renewable energy potential in the MENA region. The MENA governments would convert their cheap solar and wind electricity to hydrogen, then sell and transport via (existing) pipelines to Europe. In principle transport of energy (molecules) by pipelines is a factor 10-20 times cheaper than transporting the same amount of energy (electricity) by cables. The existing gas grids could be converted to hydrogen, which is technically easy to do and for a fraction of the cost to build a new pipeline. If this is combined with producing hydrogen from natural gas and storing the carbon dioxide in empty gas fields, a carbon free energy supply from the MENA region to Europe could be developed fast. That will all be in the mutual benefit of the MENA region and Europe.

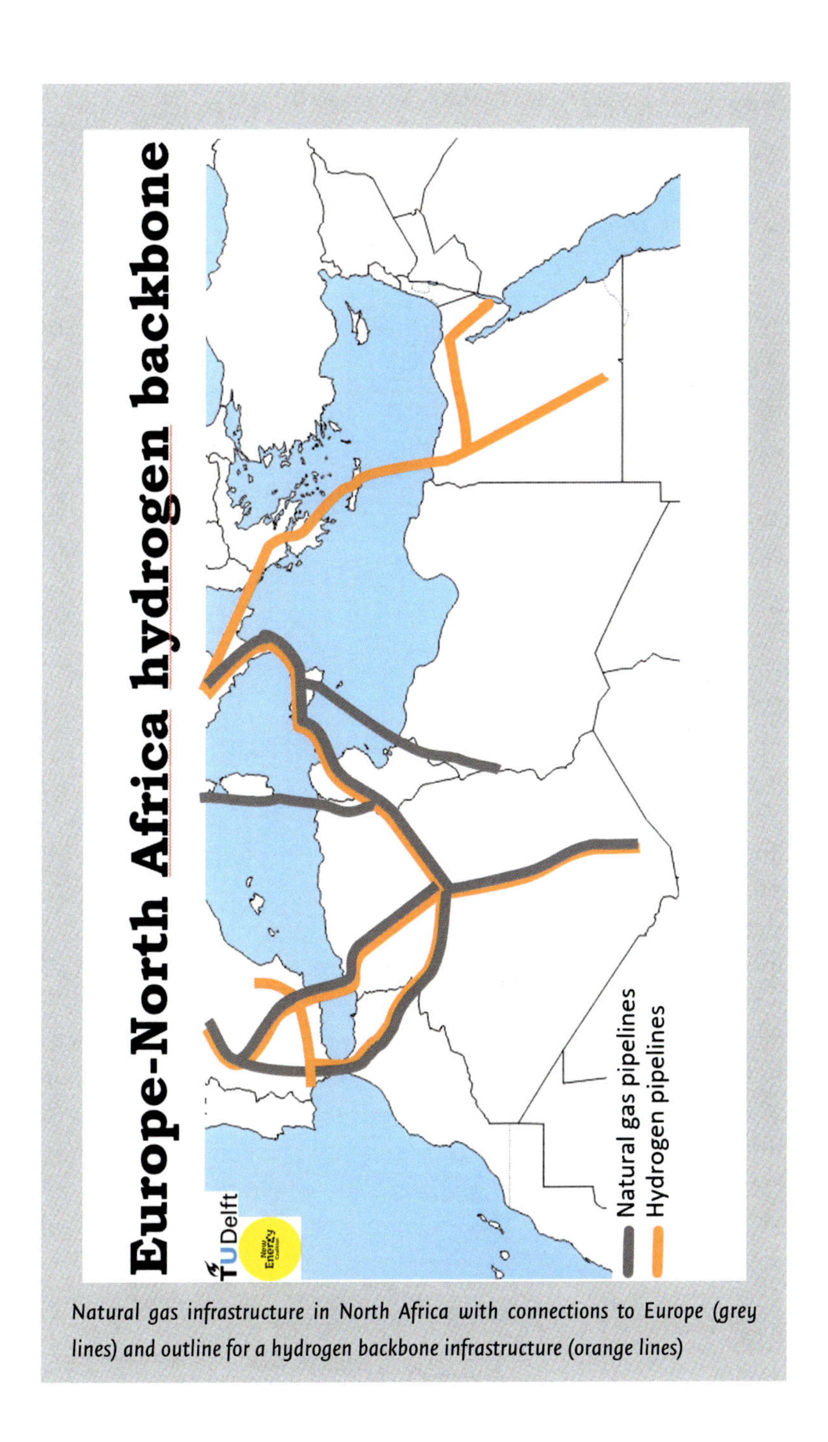

Natural gas infrastructure in North Africa with connections to Europe (grey lines) and outline for a hydrogen backbone infrastructure (orange lines)

6. Conclusion

> *'First they ignore you*
> *Then they laugh at you*
> *Then they fight you*
> *Then they join you and you win'*
> Mahatma Gandhi

What started as 'energy from the deserts', or, 'a crazy Desertec idea' has relatively quickly turned into reality. However, that occurred not without trial and error and learnings, leading to a smart prioritization of necessary steps.

How it started?
Desertec 1.0 'Electricity for Europe

In 2009, when the industrial consortium was founded, the desert power phenomenon was still completely new, sensational and imaginative: Unknown, distant deserts, the scientist Gerhard Knies with his message that only six hours of sunshine per day in the deserts would be enough to supply mankind with electricity for a year, the big industry, which wanted to support it, which was truly interested in the desert industry and promised that in a few years already the first electricity could flow to Europe and in the long run up to 15 percent of the consumption in Europe could be covered by electricity from the deserts. That was still the time of the simplified 'Desertec 1.0'. The combination of all these aspects has caused a great resonance in the international public. Together with the Fraunhofer Institute, Dii conducted various system studies and dealt intensively with potential reference projects as a training ground for the market. At the same time, the industrial network grew to include many well-known companies from all over the world. They showed a viable way for the development of emission-free energies in the deserts and their connection with Europe. Many discussions were held with experts and decisionmakers in the MENA countries. The importance of clean and safe energy for soon 500 million inhabitants in MENA increased. The future perspectives and investment possibilities were expanded by Dii, as a pioneer of desert energy, and described in detail in a series of studies.

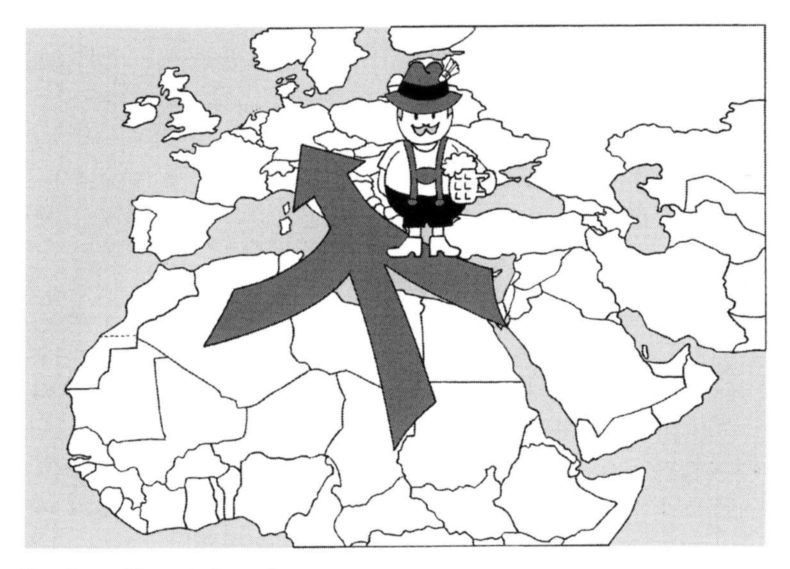

Desertec 1.0 'Power to Europe'

About four years after its foundation, the press, especially in the initiative country Germany, agreed that 'Desertec 1.0' had failed. Since there was still no electricity flowing from North Africa, the idea could not be valid. But that was a limited perception.

How it evolved?
Desertec 2.0: Renewables first for the region itself

Quickly a much more realistic 'Desertec 2.0' was formed: the primary focus on the own energy supply. The whole movement for energy from the deserts has thus come off the starting blocks faster than expected. Despite Arabellion and other turbulences, the first wind and solar power plants have already been built and commissioned in the Gulf States, Egypt, Algeria and Morocco. Not yet 'export', but 'exchange'. For many years, electricity has been exchanged without problems between Spain and Morocco and between Greece, Bulgaria and Turkey. Dii's international partners were often involved in these projects. In 2015, when Dii had just moved to Dubai, the historic upheaval came when both wind energy and solar energy became competitive without subsidies at costs below 5 eurocents per kilowatt-hour. (IRENA 2018) Since then, it has been impossible to imagine local markets without desert electricity. The expansion of decentralised and large scale emission-free energies has really gained momentum in the region, sometimes with the help of the EU, Germany and other countries and international development banks. Its feasibility is no longer doubted. Numerous concrete PV, wind and CSP projects have been developed and signal rapid growth. In 2018, PV costs will be in the range of 2 to 4 eurocents per kilowatt-hour. Producing wind power also costs 2 to 4 eurocents per kilowatt-hour and CSP plants deliver electricity for about 7 eurocents per kilowatt-hour. Dii's database for MENA and Turkey counts over 600 observed grid-connected solar and wind projects with a total capacity approaching 20 gigawatts , with a strong upward trend.

Above all local developers, but also Chinese, Japanese and some European companies are active. Although German technology, engineering consulting, insurance and financing play a role in the projects, German project developers are hardly represented.

They find it relatively difficult to deal with the risks. One of the driving forces behind Dii, ACWA Power, has managed to become the market leader in MENA. Other drivers in Dii are the CEPRI and the Global Energy Interconnection Development Corporation (GEIDCO), backed by the State Grid Corporation of China (SGCC), a prominent player in long-range intercontinental transmission networks. Innogy has made a name for itself in the smart cities sector in partnership with DEWA, and its subsidiary Belectric is building PV systems in the region in partnership with the Saudi Arabian company Al Gihaz.

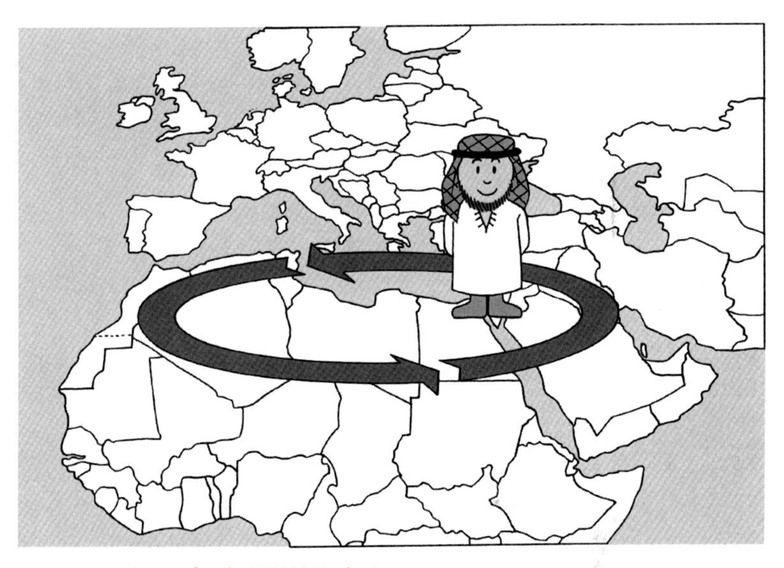

Desertec 2.0 *'Power for the MENA Region'*

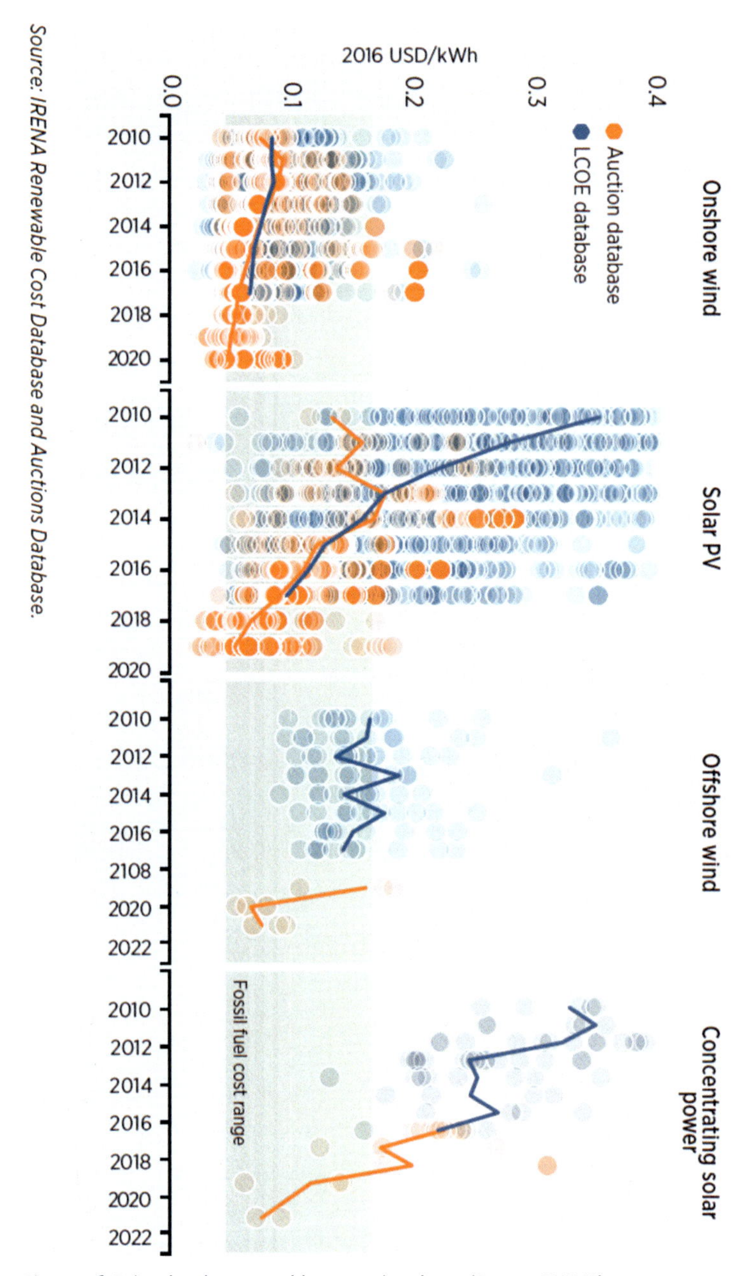

Figure 26: Prizes by the renewables are going down (Source: IRENA)

How it will go from here:
Desertec 3.0 'MENA, the global emission-free energy provider'

The overall goal of Dii was and is an emission-free energy supply in MENA and an effective energy exchange with Europe and the world. As far as electricity is concerned, the course has been set both in Europe and in MENA. The cost differences in electricity production are the driver. Exchange is therefore not a goal in itself, but a result of market conditions. The development towards a high share of emission-free energy in the energy mix and substantial electricity exchange is driven by the cost differences between Europe and North Africa and the complementary supply and load profiles. As soon as the production costs in Europe are far above the costs in MENA, the pressure to build up the grid and to transport energy in parallel will increase. CO_2 certificate prices in Europe can contribute to acceleration. As soon as the differences in electricity prices in the individual countries are sufficient, the extensive exchange of electricity can begin. This includes the extensive expansion of the network. From a European perspective, the question arises whether and how Europe would like to participate in the promotion and development of renewables in MENA and whether infrastructure expansion can be carried out in Europe and at its borders in time to enable future market-driven energy flows.

While at the beginning the focus for desert power was still rather limited to large-scale CSP plants and net electricity exports to consumption centres in Europe, Dii quickly realized that emission-free desert power is generated, transmitted, stored, controlled and consumed in all kinds of new forms, sizes and structures. The main technological role is clearly played by PV, from rooftop to gigawat scale and wind power plants that can be expanded quickly, cost-effectively and safely. Now synthetic gases or liquids, cold accumulators, batteries and other storage options come into play. CSP technology ranks third behind PV and wind.

It must prove its value in terms of controllability and storability in the local market in a context of extremely low electricity prices during the day and higher prices in the evening hours.

Figure 27: An excellent opportunity to store energy are the tanks with molten salt in CSP-power plants. Soure Thomas Isenburg

The combination and exchange of emission-free energies will lead to major synergies and cost savings, as model calculations by the Fraunhofer Institute have shown. Electricity and gas connections as well as transport ships for hydrogen etc. will be used in a new competition where they are the cheapest and safest solution in the overall context. The obvious electricity connections (often called 'supergrid') are no longer limited to the connections between MENA and Europe, also connections from the Middle East to India, Iran, Pakistan and from North Africa and

Middle East to Sub-Saharan Africa and East Africa offer interesting exchange possibilities. The 'green molecules' will find their way into world markets, just like LNG, oil or natural gas have.

Figure 28: Renewables, flexible Demand, Storage, Smart grids and Hydrogen will take away the Fossil and Nuclear Energy Industry Source: Dii

The development of desert energy in the whole area will probably not take place according to an all-encompassing timetable. The forecasts for the energy mix to be expected in 2050 have changed considerably since Gerhard Knies' first ideas about Desertec and nobody can predict today what exactly we will face. This depends on the technical, economic and political developments. However, the region knows how to deal seamlessly with the usual uncertainties. Business models are adapted to the changed conditions. Developments will, of course, initially take place locally, driven by competitive renewables, aggressive emission reductions, market impacts and government control. A sensible market organisation, the reduction of subsidies and coordinated individual planning will accelerate these developments. International cooperation provides the necessary impetus, but in principle each country will go its own way and take local interests into account. The market organisation will probably not follow the European or the Chinese model one-on-one. In MENA, all governments and market forces can and will ultimately move in the direction of renewable or emission-free energy and pursue their own ideas. A rapid increase in renewable energies naturally exerts pressure on the expansion of the required infrastructure. The electricity grids and infrastructure for water etc. will be the bottleneck in many places. More volatile generation requires substantial grid adjustments and additions.

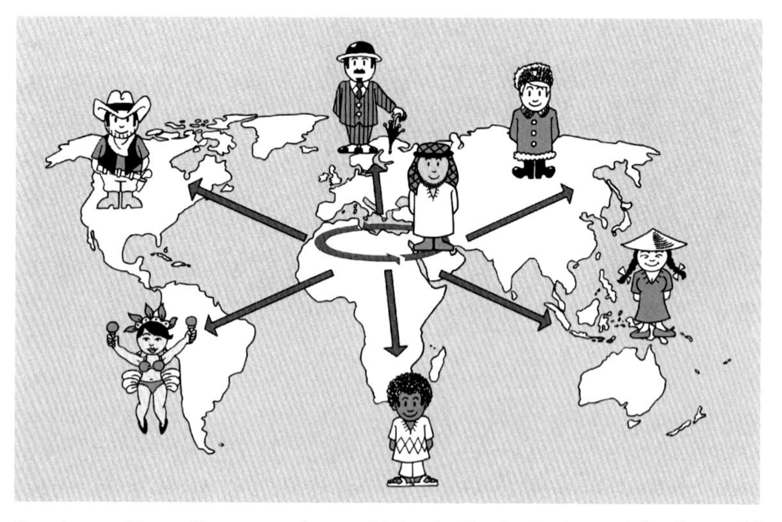

Desertec 3.0 'Green Electrons and green Molecules for the Region and for the World Markets'

Where network extensions are not possible, system-related alternatives are developed. Similar to the situation in Europe, the role of network operators will increasingly be that of a service provider for the ever more complex interplay of generation, transmission, storage and consumption. Investors in networks and network operators demand profitability from every expansion. Increasing price or cost differences between market areas should provide convincing incentives.

Desertec can no longer be seen as a 'planned economy top-down idea from a European pen', but as a 'bottom-up movement' of many locally accepted parts of the puzzle, with benefits for the local population and industries.

Of course, the established oil and gas industry is not happy about the disruptive decarbonization. This may explain why MENA initially lagged behind in the implementation of renewable energies. First of all, the traditional energy importing countries, Morocco at the forefront, took the initiative. In the meantime, however, all MENA governments have understood what is at stake through the COP21 Paris Climate Agreement and the competitiveness of renewables. The oil and gas exporting countries, such as Saudi Arabia, the United Arab Emirates and Algeria, have ambitious plans and are implementing them. Desert energy is no longer a mirage but is being perceived positively by more and more market participants. The region is just discovering the great forces and synergies that are at work in energy exchange and export. The integration of renewables into the networks, linking markets for green electrons and green molecules, storage, seawater desalination, energy-intensive industry, production of technical products and solutions, electric vehicles, smart cities, smart office, smart home and responsive demand are the technical topics that the region has to deal with. Dynamic price signals, green capacity auctions, green priority access, emission cap and trade, green certificates, green customer products etc. will further accelerate the energy transition. There are certainly still many practical issues to be resolved, but system studies and practical examples in other markets show how emission-free energy supply can and does work.

> *'A swift energy transition is a matter of connecting private and public stakeholders with effective emission penalties market-based incentives for companies and benefits for the region'*

The energy transition will reduce dependence on imports, lead to new business, create jobs and thus contribute to geo-political stability. Stability inspires confidence and enables investment from the private sector. Local companies, private companies such as ACWA Power (Saudi Arabia) or the agency MASEN (Morocco), DEWA (Dubai) and Masdar (Abu Dhabi) are already showing that the construction of large solar and wind power plants or programmes for PV roof systems is not being driven by Europe, Japan, China or the USA, but by the countries in the region themselves. The energy transition will grow rapidly within the region and intercontinentally from the first local beginnings that we are now observing. It connects governments, industry, institutions and people. MENA has the potential to become an emission-free powerhouse. Commercial capital in principle has to compete with sectors and regions worldwide. The business cases in the projects in the various MENA countries are not among the most attractive worldwide, but the rapid development nevertheless shows that investments are made everywhere. The increased cooperation between international initiatives and authorities has meanwhile become a guarantee that desert energy is professionally supported in the private and public sectors. The process is unstoppable. After the necessary preliminary work has been successfully carried out and the caravan has been set in motion, even the most cautious developers and investors may have an appetite for project development in MENA.

Part 2 DEEP DIVE Information

A lifely network of Associated Partners

By Cornelius Matthes, Senior Vice President Dii Desert

In ten years time the industry network of Dii has developed into a unique mix of companies with a clear, longterm footprint in MENA and companies that are just setting foot in

Cornelius Matthes is sitting on the right. Source: Dii

this challenging market. With pleasure I highlight my experience with selected companies, many of which we have helped to find their way in the MENA market:

Terna Energy (Greece), is a leading IPP in central Eastern Europe, with a large Wind Project in the USA. Terna Enegy took the initiative to host one of the first Associated Partner Meetings' in Athens, with a memorable surrounding of the Akropolis Museum. Terna Energy provided exceptional contributions in the Dii Working Group on Wind Projects. The company later became the first Associated Partner that was 'promoted' to a full shareholder.

IntesaSanpaolo competed with **UniCredit** to become the first Italian bank as a Dii shareholder. Ultimately, UniCredit succeeded but IntesaSanpaolo with their italian generosity and good style became one of the most active Associated Partners .
For quite some years Italy was the second strongest country represented through its industry in Dii, almost outnumbering German companies. **Enel Green Power, Terna, Unicredito, Intesa Sanpaolo and Italgen** were quite active. **Italgen** is a leading IPP for hydropower, with a strong expertise in wind parks, owning a sizable asset in Bulgaria and developing wind farms in Morocco and Egypt for own consumption to cement plants of Italcementi.
MauriSolaire, a small renewables developer in Mauretania has been among the first Associated Partners. We are prou dthat it is still is part of the Dii group in 2019. The owner, Mohamed Lemrabott El Moctar has shown outstanding personal engagement for the Dii mission over 10 years.
Another long standing supporter is the German/Austrian engineering consultancy company **ILF** with helpful engineering and operational support.
While some companies tried to put the focus on promoting own products and services in the whole network of Dii, others acted differently and improved their business smartly indirectly leveraging on Dii's contacts and credibility.
ABB has been a founding shareholder of Dii and Associated Partner since Dii moved to Dubai. The company has been contributing immensely to Dii's work, in different fields from power grids (with specific focus on HVDC technology) to storage or different technology for solar PV systems.
Saudi group **Al Gihaz** has been an Associated Partner since 2017. The group has been a pioneer in new market segments of solar PV, They are the leading shareholder of Belectric Gulf, the local arem of associated Patner Belectric.

Amana of the UAE joined Dii also as an Associated Partner in 2017. The construction group which has the highest market share for district cooling (a strategic field as district cooling will play a key role in load shifting based on cold storage) has been among the first to set up a solar PV division to engage on the shams Dubai program for net metering.
E-nara, a leading Egyptian developer and IPP of solar PV projects, has been part of the Dii family since some time. In addition to good contributions, E-nara sperheaded the initiative to organise a first major event to inaugurate the construction start of the 1,500 MW solar PV park in Benban in upper Egypt, with the presence of the Governor of Aswan, and top officials from the federal government, utility and development banks financing the solar park.
Europagrid of Ireland has a unique profile among the Associated Partners of Dii, with the core business in developing electrical interconnections with a merchaent model. Dii and Europagrid have been working together for a feasibility study on a subsea cable between Tunisia and Italy in the context of a EU Ten-E grant. More recently, the group is also active in electricity trading and new business models.
Enerwhere joined Dii in 2019. This young company has uniquely pioneered the market for mobile and permanent hybrid projects (diesel/PV/battery) on an IPP basis, with short term power purchase agreements. The company has also built the second largest solar PV rooftop globally (18 MW) on the 'Mai Dubai water bottling factory and water reservoir.
Worley Parson joined Dii in 2017 and has made significant contributions in the field of green molecules and hydrogen, a topic Dii has recently launched a working group putting emphasis during the work from 2019.
Yellow Door Energy is also a new Associated Partner of Dii and a pioneer in the field of industrial/commercial solar

PV projects with private power purchase agreement and full energy service provider to such clients. The company recently raised USD 65m from a prestigious group of international investors such as IFC, Mitsui, Equinor and Apicorp. **Navigant**, formerly among others Ecofys, has been closely involved in Dii since 2017, and more recently was awarded the project to manage the German UAE Energy Partnership. **Krinner** joined Dii only in 2018, but made big headlines, building the screw foundations and mounting systems of world's largest solar park, the Sweihan 1,200 MW park in Abu Dhabi. They managed to do that in record time, and mostly with automated robots, working 24h and controlled by satelite from their home base in lower Bavaria in Germany.

Among the several Associated Partners joining in 2019, there is also Italian EPC company **Saipem** which publicly announced to aim for a transition to half of the revenues from green projects in the mid term which is a massive change for a group today still dominated by oil & gas. Also, **SQM** from Chile as world leading provider of malten salt for thermal storage in CSP plants came on board, to leverage on Dii's track record in CSP and develop business.

In addition, **Envirofina**, an innovative company based in Dubai which has been pioneering the ESCO market segment, devering the largest ever retrofit projects with LED lamps, delivering more than 500,000 LED lamps for Carrefour markets in 12 countries. Now, the company is a front runner in outdoor ESCO projects with integrated street lighting.

The newest addition to the group is **Ecolog**, a leading international group, providing all kinds of service to clients in remote locations and often difficult countries. Dii Desert Energy is delighted to help in advising on how to make specifically energy service more emmission free.

1. Country examples

From Dubai, Dii observes the development of renewable energies in MENA. In 2017, 496 projects were counted in the region:

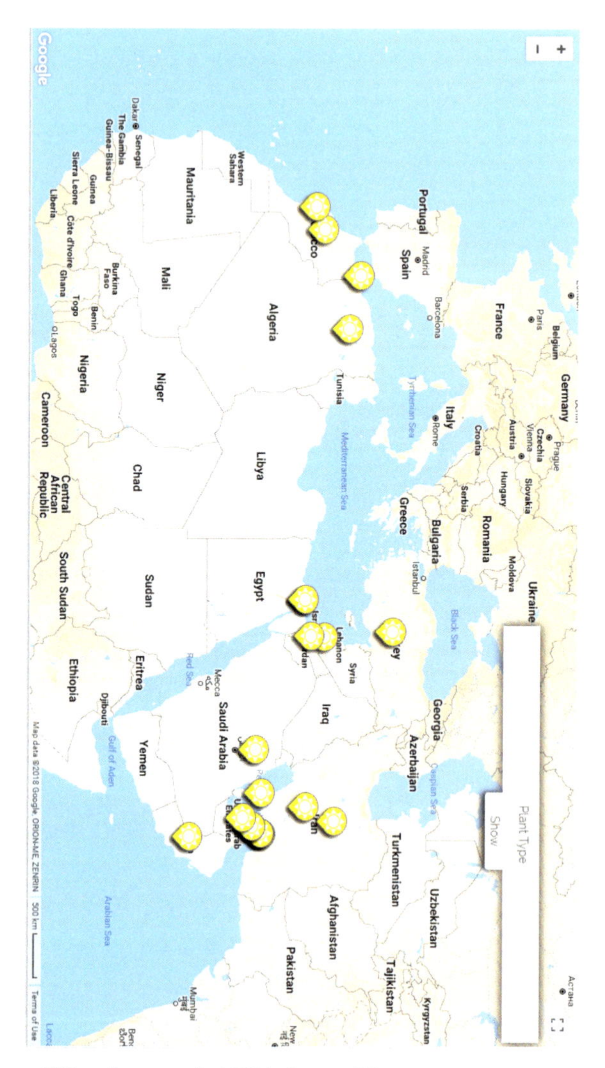

Figure 29: 24 CSP projects 2017 in MENA. Source: Dii

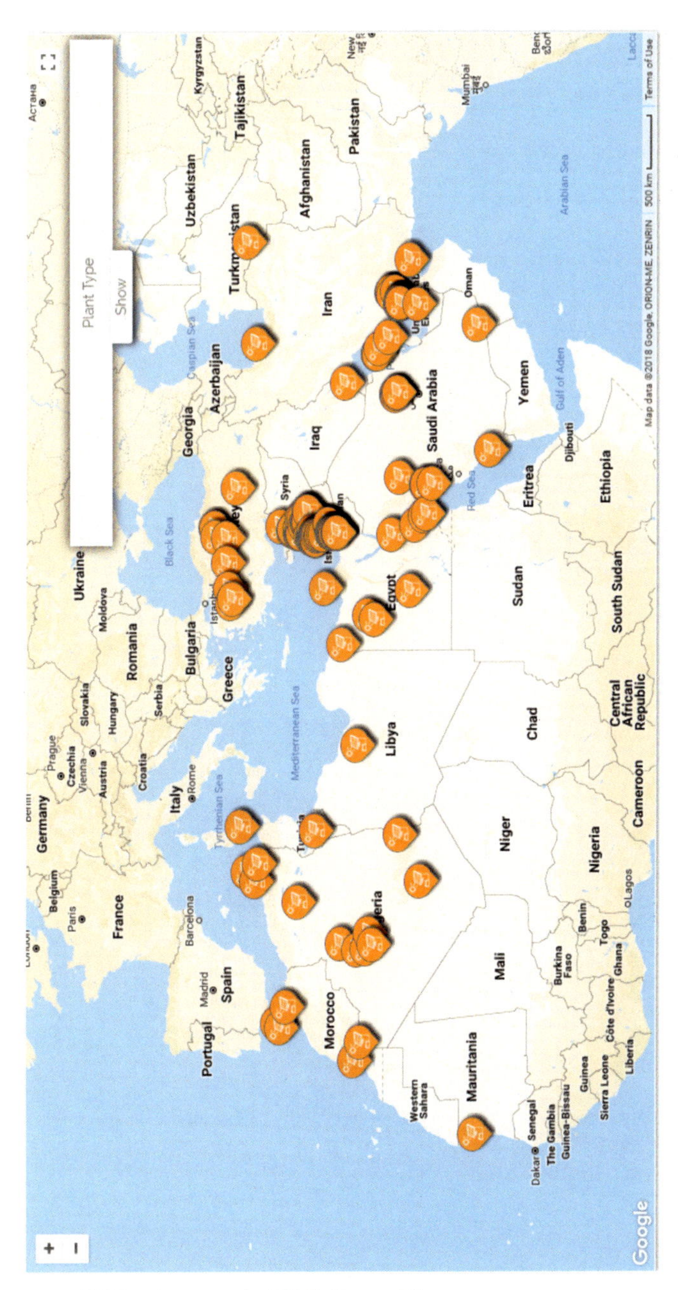

Figure 30: 256 PV projects 2017 in MENA. Source: Dii

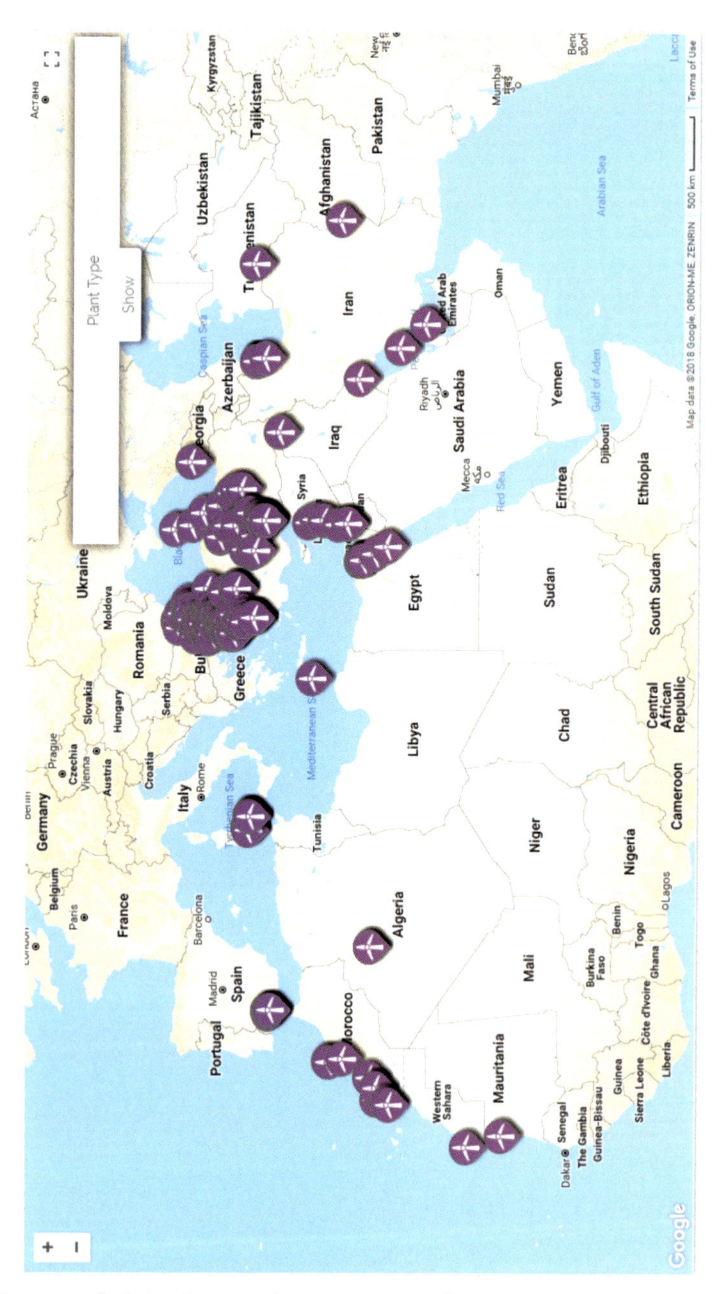

Figure 31: 216 wind projects 2017 in MENA. Source: Dii

1.1 Pioneer Morocco

The concrete implementation of the Desertec idea is most visible in Morocco. Large and medium-sized power plants as well as small rooftop PV plants are growing like mushrooms.

Morocco is the most north-western country in Africa with a geostrategically important position on the Strait of Gibraltar. In Tangier and Casablanca there are tax-privileged free trade zones around the major ports. Since July 1999, King Mohammed VI has been leading to the country. Under his reign, the country's economy (with 35 million inhabitants) grows by 5 percent annually, but the social differences are still large. Since the King claims to be a descendant of the prophet Mohammed, he also assumes a spiritual leadership role. The King enjoys a certain popularity with his subjects. But every now and then there is unrest in his Kingdom. The monarch reacted to the Arab spring with cautious reforms. The 2011 Constitution provides for a constitutional monarchy with the elements of parliamentary democracy. For this purpose, a referendum was held in which the King, 47 years old at the time, ceded part of his absolute power. Now the Prime Minister is elected by the people's representatives, no longer by the King. The Prime Minister's power was also strengthened because he appointed the members of his cabinet himself. However, the monarch must agree to all personnel proposals. Many of his subjects enjoy the new freedom, people from other Arab countries flock to Morocco. When Mohammed VI took office, he surrounded himself with a number of men – among them Bakkoury – who were supposed to help him transform Morocco into a progressive country with a flourishing economy.

The country is well placed for investment thanks to political stability and steady economic growth. However, unlike its neighbour Algeria, Morocco has no fossil raw materials and is dependent on energy imports from hard coal, natural gas and oil (DIHK 2016). Morocco had a total power plant capacity of 8,000 megawatts in 2015. Hard coal accounted for 50 percent of this. It is mainly imported from South Africa. Morocco's largest coal-fired power plant is located 130 kilometres south of Casablanca. It is owned by TAQA, of the United Arab Emirates, and has a capacity of 2,056 megawatts. In recent years, Morocco's electricity consumption has risen by 7 percent per year as the country industrializes and electrifies itself. In recent decades, many, often remote villages have been supplied with electricity for the first time. From 1995 to 2010, their electrification rose from 20 to 96.8 percent. Morocco is still dependent on imports of electricity from Spain because of its ever-increasing energy needs. Even in 2015, 5,833.70 gigawatt-hours were imported from the Iberian Peninsula – 18.5 percent of the total energy demand. To ensure its energy supply, Morocco is linked to Algeria and Spain via high-voltage lines. The capacity of the connection with Spain was doubled in 2006 from 700 to 1,400 megawatts. There are also plans for power lines to sub-Saharan Africa.

The country in the Maghreb wants to disconnect itself from this supply of primary energy, which swallows foreign currencies and pollutes the climate. In view of the expected increase in energy demand, power plant capacities are to be expanded to 14,500 megawatts by 2020, of which 2,000 megawatts will be generated by solar, wind and hydropower. That is why the country is going through an energy transition. By 2020, 42 percent of the electricity generation capacity is to consist of renewable energies – one third each of solar and wind energy and hydropower.

Since 2010 MASEN, with Bakkoury at the top, is in charge of the implementation in the field of solar energy, prepares projects and implements tenders. MASEN's task is to implement renewable energies of the highest standard in Morocco and to build up national expertise as far as possible. In addition, there is economic, ecological and social support for the affected regions. The King attaches a high priority to this, which has made Morocco, along with South Africa, one of the pioneers of renewable energies in Africa.

Morocco's commitment is supported internationally. In April 2016, the German Vice Chancellor and Federal Minister of Economics and Energy, Sigmar Gabriel, visited the country. He was enthusiastic about developments in the energy sector: "I don't know of any other country that is driving its energy policy forward with such precision, clarity and speed of implementation as Morocco" (dpa 19.04.2016). The following November, Marrakech hosted the COP22 World Climate Change Conference. However, the conference was overshadowed by the threats of the newly elected US President Donald Trump to withdraw from the climate protection agreement. A step he actually took in 2017, that isolated the USA and triggered a strong movement in the country (Ehlerding 01.06.2018).

Nevertheless, the Moroccans proudly demonstrated their progress in the construction of the Noor power plant in Ouarzazarte. During the COP22, numerous conference topics found their way to the place that is significant in terms of technical history. Federal Environment Minister Barbara Hendricks, accompanied by journalists, also embarked on the 200-kilometre journey from Marrakech to the rapidly expanding provincial city. This trip takes you through a beautiful landscape and leads via winding roads over the 2,260-meter high Tizi n'Tichka Pass. The route to the power plant complex is one of the main connecting roads over the High Atlas. In winter it is not seldom closed because of heavy snowfall. Along the way there are numerous Berber villages, which give the viewer an insight into the simple agricultural life of the region. In Ouarzazarte the solar power plant complex on 3,000 hectares is under construction. The parabolic trough power plant Noor I has been producing electricity since February 2016. A salt reservoir can store the energy for three

Figure 32: The CSP tower in the solar power plant in Ouarzazarte. Source: Thomas Isenburg

Altogether, the expansion will take place in four stages and would be completed by 2018. Noor II is also a parabolic trough power plant with a capacity of 200 MW, Noor III is planned as a tower power plant with a capacity of 150 MW. The tower is the highest in Africa. A total capacity of 510 MW will thus be available from CSP power plants. Since water is a scarce commodity in the desert, air cooling is planned for Noor II and III. The prices for the electricity produced have fallen by 30 percent to around 11 eurocents per kWh. In Morocco, electricity from photovoltaic systems can be obtained for about 3 eurocents per kWh with favourable solar radiation. However, they do not provide any electrical energy in the dark.

Local content is at least as important as a secure, clean and affordable energy supply. In August 2016, about 4,000 people were working on the power plant complex. Many came from the surrounding area. Together with Dii shareholder ACWA Power, MASEN is implementing various social projects such as the cultivation of date palms in the immediate vicinity of the huge facilities. The Moroccans take on tasks such as electrical installations, steel structures, safety, cleaning and logistics. Whereas in Noor I only 30 percent of companies were from Morocco, in Noor II and III the figure is already 50 percent, i.e. more than 100 Moroccan companies. One goal of the Moroccan government is to implement as many value-added processes as possible in its own country in all industrial sectors in order to boost local markets. A separate research infrastructure was also set up. Most of the financing comes from Europe, even Germany, the technology is from China and project developers are Saudis and Moroccans. The cooperation between the Chinese and Moroccan subcontractors is not always easy, the construction time is prolonged and the quality of the work also suffers because the local workers still have little experience.

We asked Mustapha Bakkoury, the founding CEO of MASEN, about his vision on the energy future and job creation in Morocco

Mustapha Bakkoury. Source: masen

Recent renewables capacity in Morocco amounts to a few Gigawatt. In addition to electricity production from renewable sources, the purpose of MASEN's integrated approach is to establish strong and competitive renewable energy

ecosystems. MASEN is striving to create a competitive economic fabric that efficiently harnesses existing skills and helps create new ones.
The development of a competitive, local sub-sector involves the intensification of industrial integration. It is about optimizing the contribution of local manufacturers to a given economic sector. Our goal is to promote a vertical integration to create benefit at the local level, for that matter, industrial integration is destined for renewable-energy companies based in Morocco. We endeavour to reduce entry barriers, and are committed to the development of a competitive industrial ecosystem, through contributing to the inception of large-scale energy projects, while the cluster helps in the development of a national industrial fabric capable of supporting this dynamic. The projects set up by MASEN also create thousands of jobs during their construction, and allow the integration of national players, which allows the rise in competence of the national human resources. During construction, the total number of employees (at the peak) per project was the following:

- Noor Ouarzazate I: over 2,000
- Noor Ouarzazate II: more than 4,000
- Noor Ouarzazate III: more than 2,500

The local population is taken into account, upstream and downstream.
For the Operation and Maintenance (O&M) phase, MASEN has launched, with two of its partners, a training programme which aims at creating operators and technician position. The programme, launched nearly a year ago, has already trained 100 students, of which 60 found a position in the O&M of Noor Ouarzazate II and III. Also, it is important to note that this programme helps to transfer knowledge to the local training centres and trainees.

How and why is Morocco leading the way in solar energy development, and how is it pioneering climate change technology?

In terms of technology, the national strategy aims to optimize the domestic technological mix in order to implement a set-up that can provide a stable, flexible and smart system capable of incorporating an increasing share of clean energy. In terms of innovation, MASEN provides industrial projects with applied and pre-operational research and development expertise in order to nurture the renewables ecosystem and make renewable energy projects increasingly efficient, accessible and affordable. As part of our work, MASEN is fast-tracking the emergence of innovative green technology on the market. MASEN'S R&D strategy involves identifying developing technology with high potential and supporting it through the demo and pre-production phases. MASEN R&D achieves this with the support of Morocco's academic and innovation ecosystem, for which it organizes various training and support incentives, such as the MASEN Talents Campus and the MASEN Talents Awards. Finally, thanks to the Ouarzazate R&D platform, a 200-hectare-plus site designed to accommodate pilot projects, MASEN R&D can offer domestic and international green technology leaders and developers the opportunity to test their innovative systems to scale in real operating conditions. This valuable proposition has already attracted various multinational groups and start-ups who have formed partnerships with MASEN R&D.
MASEN is also open to new forms of clean energy (tidal, waste-to-energy, electric mobility, etc.).
In terms of renewables, the future of Morocco is reaching at least 3,000 MW of installed capacity by 2020, and at least 6,000 MW of renewable installed capacity by 2030, which will cover over 52% of Moroccan energy needs.

Dust on the mirrors of solar thermal power plants, which critics call the main problem, is actually a topic. There is a lot of dust in Ouarzazate. It disturbs the reflection of sunlight. The solution is well-coordinated cleaning cycles. The complex steam cycle in the desert is more of a headache, as it must function in a stable manner for decades in order to operate efficiently.

The successful construction and operation of the plants shows that a good starting position has been created so that CSP can assert itself in the long term in the energy transition. The new drivers in this development, DEWA and ACWA Power, set a new cost record of approximately 7 US dollarcents per kWh in 2018. In Morocco, further construction projects are planned in the towns of El Aaiún, Boujdour, Tata and Midelt. Here, a hybrid construction method consisting of CSP and PV power plants is primarily to be implemented. The PV systems can cover the peak consumption during the day. At night, the CSP power plants that can store energy can then supply the electricity in the event of supply bottlenecks. An other solarcomplex will start in the Middelt a small town between the high and middle Atlas although with the name Noor. As part of the development of the Noor Midelt solar complex, and following the tender of Noor Midelt I plant, Masen announced the launch of the Request for Qualification for the solar plant Noor Midelt II in July 2019. The Noor Midelt solar project will be made up of two solar plants which will combine thermosolar and photovoltaic technology to produce 800 MW of renewable energy, with a storage capacity of 5 hours. The hybrid technology will optimize energy production, leading to competitive pricing (e.g. 6,3 Eurocents per kwh at peak time or even lower).

The expansion of wind energy in Morocco has also been making considerable progress for about ten years now, as high wind speeds prevail on the Atlantic and Mediterranean coasts in particular. These are between eight and eleven meters per second at a height of 40 meters, an attractive value for onshore systems. Experts calculated from this a capacity of 25 GW. The windy coasts of Morocco have a length of 3,500 kilometres. Large wind farms are already located in Essaouira, Tangier and Tétouan. One advantage is the low population density in Morocco, compared to Germany, which averages 80 inhabitants per square kilometre. The Federal Ministry for Economic Cooperation and Development is supporting the Moroccan government's integrated wind programme with 130 million euros. Three wind farms with a total installed capacity of 450 MW are to be erected in northern Morocco. In private projects a price of 2.5 eurocents per KWh is achieved. Capacities for expansion are also approaching the 2,000 MW target set for 2020.

Figure 33: Wind farm near Tangier. Source: Thomas Isenburg

An 850 MW wind farm in Tangier accounts for a large share of this. The German technology group Siemens, the Italian project developer Enel Green Power and the Moroccan energy company Nareva, all former shareholders of Dii, won the contract. It involves an investment sum of 1 billion euros. The project involves the construction of five wind farms and a rotor blade factory – the first on the African continent. The factory was opened in October 2017 by Moulay Hafid Elalamy, Minister of Industry, Investment, Trade and Digital Economy, and Siemens Gamesa CEO Markus Tacke. A capacity of 4,300 MW is to be installed in Morocco in the coming years – slightly less than the capacity installed in Germany in 2016.

Figure 34: Opening of the first rotor blade factory on African soil by the Minister of Industry, Investment, Commerce and Digital Economy Moulay Hafid Elalamy and Markus Tacke in Tangier. Source: Thomas Isenburg

Only a handful of Europeans work in Tangier, and half of the Moroccan employees were trained at the Siemens Gamesa site in Aalborg, Denmark, for the sophisticated process of manufacturing rotor blades, which are exported from Tangier to the Middle East and southern Europe. The companies benefit from the free trade zones around the ports of Casablanca and Tangier.

Another interesting example of local production is the Moroccan presence of the German cable manufacturer Leoni. In four plants, the group employs approximately 8,000 people in the country. Leoni played a key role in the Dii study on local value creation.

Travel notes from Thomas Isenburg: A risky research trip to Morocco

Thomas Isenburg. Source: privat

The journalist Isenburg has been involved in the topic of sustainability both privately and professionally since the Nobel Prize was awarded to Al Gore and the IPCC. He made initial contacts with the Club of Rome. With the media hype around Dii he dedicated himself to the topic Desertec and planned a first research trip to Morocco to be able to report authentically. Isenburg was the first German journalist to write an illustrated report about the construction site of the solar power plant in Quazarzate (Isenburg 27.01.2014). He frequently travelled to North Africa and spoke with international experts. He visited the Noor project in Ouarzazarte at regular intervals, including in April 2015, but this time he was to have unpleasant experiences regarding freedom of the press in Morocco.

Diary entry from April 2015:
First, everything goes according to plan, I fly with Ryanair from the airport Düsseldorf Weeze directly to Marrakech. Only half of the seats on the plane are occupied, I pick up many scraps of French conversation. The passengers reflect Moroccan society well, as I know from my previous visits.

Also, the first hours in Marrakech go as usual. I take the bus to the Medina and take a room in the Mimosa Hotel, a simple small hotel with many Moroccan guests. In the medina a lively hustle and bustle, the great social differences in the country are immediately visible. The next morning, we go to the bus station in Marrakech. Bus travel throughout the country is widespread in Morocco. The winding roads and exhausting drive through the Atlas to Ouarzazate takes about 5 hours. Many insights into rural life are possible. Once arrived in the small town, I take a room at the central square of the city. All of life here takes place in the evening. There is also the car rental of Mohammed, where I rent a car with driver, as usual. Mohammed is about 50 years old, has studied English and has worked for the GIZ before. He knows the region like the back of his hand and instructs the driver to visit the power station. When we get there, I start taking pictures. I get off at the gate to the Noor II power station complex, around which a fence has been built. Young Moroccans demonstrate in front of the construction site. A policeman comes up to me and forbids me to take pictures. I'm going back to the car. Meanwhile he has called in an English-speaking policeman and confiscated my camera. Two more policemen arrive, with more and more stars on their epaulets. Finally, a commissioner is requested. I'm starting to realize I'm in trouble. My driver looks afraid. He doesn't have the necessary license that a tour guide needs in Morocco. We're being driven to the police station. They lead us into a room with a cell and start interrogating us. After about two hours, a Commissioner appears and asks questions. The cop wants to know why I took the pictures, and I'm explaining my mission. After another hour, the driver and I are taken to another room and interrogated by five or six police officers. They also want to know the reasons for my frequent trips to Morocco. We're left alone. After a while, another man comes to us and reassures us. He shows interest in the project and my travels to South Africa. I get my camera back – but without the memory card. I have to sign a long protocol in which I declare that the limits of reporting will be respected. I had not obtained the necessary permissions – a mistake that would not happen to me again. During further visits, I secured in advance the support of MASEN, who gladly showed the complex in the environment of the COP21. We'll be released in the evening. Mohammed and I sit together for a long time over tea and talk about what we have experienced. He explains to me that the police in Morocco have a high standard, the police officers are partly trained in the USA and are very present in everyday life. The demonstrators demonstrated in search of work – the construction of the power plant had hardly had any positive effects on the local labour market. Several thousand people are to work at the power plant complex, many of them from China or Spain. Energy is not an issue for most Moroccans. The Moroccans want good medical care and an efficient education system, explains Mohammed, who is an eager newspaper reader. The public hardly notices the project. Mohammed and I will be friends. We also talk into the night about the similarities between Christians and Muslims. The educated Moroccan advises me to stay on the subject of electricity from the desert.

Back home again, Isenburg learns more about Morocco from Reporters Without Borders:

"Criticism of the King is forbidden there and has already been punished several times with prison as an attack on the nation's sacred values. Western Sahara policy and the corruption of high-ranking politicians are also taboo topics. The most important television and radio stations are controlled by the state. Uncomfortable journalists are sometimes charged with crimes (e.g. drug offences). The intimidation repertoire also includes ad boycotts, threats, character assassination, assaults and burglaries. Foreign media are sporadically banned or their accreditations withdrawn from their correspondents. Morocco ranks 135th out of 180 in the Freedom of the Press ranking."

1.2 *Algeria*

Algeria is the most populous of the three Maghreb countries and the largest on the African continent. The Sahara occupies about 80 percent of its area. The approximately 40 million Algerians have the highest per capita income in Africa. Unlike Morocco and Tunisia, Algeria has large gas and oil reserves. These meet the country's growing energy needs and their export earnings support the state budget. Algerian natural gas enters the EU through a pipeline running from Morocco under the Strait of Gibraltar to Spain. Another pipeline leads via Tunisia to Italy. Oil reserves are estimated at 12.2 billion barrels. However, prices fell in the second half of 2014. Reduced profits led to a lack of funds for housing construction and infrastructure. The latter was no longer able to keep pace with the growth of the gross domestic product of three percent.

In recent decades, Algeria has been unstable. The country underwent the transition from a planned economy, Algerian socialism, to a free market economy. The legacy of the former planned economy is extensive bureaucracy and widespread corruption. The uncertain political and social atmosphere will be exacerbated in 2018 by President Abdelaziz Bouteflika's years of critical health. Bouteflika has only appeared sparsely in public in recent years. Disillusionment and disenchantment with politics among the population are also reflected in an extremely low turnout. The themes in Algeria are the fight against terrorism and the relationship to the former colonial power France. Conditions that do not exactly attract private companies and foreign investors. Nevertheless, the Algerian government is increasingly trying to focus on renewable energies. It also knows that the fossil age is slowly coming to an end and wants to reduce the consumption of fossil energies. Algeria signed the Kyoto Protocol in 2005.

Energy history has already been written in Algeria. The power plant Hassi R'Mel, in operation since 2011, is considered to be the world's first solar hybrid power plant. It consists of a solar thermal and a conventional gas and steam combined cycle power plant. The power plant has a total capacity of 150 megawatts, of which 30 megawatts can be provided by solar energy. Another special feature is the first officially fixed feed-in tariff in the region. It was adopted in May 2014. Plants with an installed power generation capacity of at least one megawatt will receive support for twenty years. The electricity supplier Sonelgaz pays the guaranteed purchase price and gets back the additional costs from the national fund for renewable energies in return. However, fossil fuels are still relevant. Critical forces in the government have so far been able to effectively slow down the implementation of further plans for feed-in tariffs. Individual PV projects throughout the country deliver a total output of 233 megawatts, of which the major part was awarded to a Chinese consortium in early 2015 as EPC contracts (construction of turnkey plants). However, 85 MW were built by the German EPC contractor Belectric (which was taken over by Innogy in 2017). The plants were built between 2015 and 2017 and made Algeria the country with the largest installed capacity of PV power plants in the entire MENA region by 2018. It is owned and operated by the Sonelgaz subsidiary SKTM.

Algeria wants to implement 22 gigawatts of generation capacity for renewable electricity by 2030. 12 gigawatts are for its own use, 10 gigawatts for export. This corresponds to 40 percent of Algeria's energy requirements being covered by alternative energy sources. The energy mix for renewable energies according to the ideas of the Algerian government is as follows: 15.5 gigawatts photovoltaic, 5 gigawatts wind energy, 2 gigawatts solar thermal power plants, 1 gigawatt bioenergy, 400 megawatts combined heat and power plants and 15 megawatts geothermal plants. Algeria has created its own Ministry for Renewable Energies to promote the programmes. The new ministry began its work after the parliamentary elections in August 2017 and is a clear signal from the Algerian government. The oil and gas industry no longer offers prospects, in particular for the 65 percent of Algerians who are under 30 years old. For many years, experts have pointed out that Algeria's focus on the export of oil and gas has no future.

The Algerian development programme for renewable energies and energy efficiency is divided into two phases. In the first phase until 2020, an output of 4,000 MW consisting of photovoltaics and wind power is to be installed. The lion's share of capacities for renewable energies will then be built between 2020 and 2030. However, there are always difficulties in the preparation of tenders. Interestingly, four local factories for the production (assembly only, no cell production) of PV modules have been built in recent years. However, annual production capacity is low and bankability rather questionable. One of the main obstacles to the development of renewable energy in Algeria is the low market price of electricity. Private customers pay 4.4 eurocents per kilowatt-hour. For industrial customers, the price is 2.2 eurocents per kilowatt-hour.

In May 2018, Dii organized a conference in Algiers in partnership with the local solar association Club des Energies. The event was attended by high-ranking representatives of the government and the state utility Sonelgaz. The aim was a dialogue between the developers, (promotional) banks and the government in order to complete the first tenders. For some time now, there have been intensive discussions, including with the IFC (International Finance Corporation, part of the World Bank Group), about the requirements of the PPA (long-term power purchase agreement) in order to make the projects attractive for international developers and investors.

In July 2018, in fact, the first tender for a total of 400 megawatts was announced. In spring 2019 the political situation in Algeria seems to change. Protests among the population lead to the fact that the frail President Bouteflika no longer stands for election but wants to initiate reforms. Supporters of the protests are job seekers, trade unionists, feminists, journalists and opposition groups. In 2019, Bouteflika's health is rapidly deteriorating and he will only function as the marionette of an obscure power apparatus. This is dominated by the military and still goes back to the bloody liberation war against the colonial power France from 1954 to 1962. This withdrawal testifies less to a new 'political spring' than to the fact that North Africa and the Middle East are experiencing an ongoing process of tenacious change. Unemployment is very high almost everywhere in the region. In some countries in North Africa and the Middle East, around a third of young people are unemployed. In spring 2019 demonstrations in Algeria started again. The mostly young demonstrators demanded the Resignation of the 82 years old president Bouteflika. Than he stepped back in april 2019. The young algeriens are still waiting for reforms. But the powerful persons and generals from the war against France are still control the army and the police.

1.3 *Tunisia*

Tunisia, the cradle of the Arab Spring, now has the most progressive constitution in the Arab world. The Nobel Peace Prize was awarded in 2015 to the dialogue quartet from the Tunisian Trade Union Confederation (UGTT), the Employers' Association (UTICA), the Human Rights League (LTDH) and the Bar Association.

Tunisia is characterised by relative stability. The emergence of a new democratic constitution was a decisive step for the country's economy. This resulted in great sales potential in the solar energy sector. Tunisia also has few fossil resources compared with its neighbouring countries. There is also an energy deficit. The well-developed power grid connects Tunisia with its neighbours Libya and Algeria. However, structural exchanges of electricity have so far hardly taken place via these connections. An electricity connection to Italy has been under discussion for a long time. Algeria also exports gas to Italy via Tunisia.

According to the ideas of those responsible, 30 percent of the electricity fed into the Tunisian grid is to be generated from renewable energies by 2030. The aim is to build PV and wind power plants with a share of 50 percent each. The programme is expected to create more than 30,000 urgently needed jobs in parallel.

The potential for wind turbines for onshore and offshore wind farms on the Tunisian coast is estimated at around 10 gigawatts. Starting in 2007, only a few wind turbines in the form of small parks with outputs of between 10 and 45 megawatts were initially installed. The Metline and Kchabta wind farms near Tunis, with a total capacity of 190 megawatts, were thus built.

Figure 35: One of two larger wind farms near Tunis in Tunisia. Source: Thomas Isenburg

After the successful commissioning, 26 generators were added. Today, the electricity is fed into the Tunisian grid. According to the expansion plans of the Tunisian authorities, the target is 1,750 megawatts by 2030.

The expansion of solar energy is being driven forward under the name Pro Solar. The 30/30 strategy of the national energy agency ANME (Agence Nationale pour la Maîtrise) stipulates that by 2030

30 percent of energy will come from renewables. The approximately 3,000 hours of sunshine per year at a radiation density of 1,850 kilowatt-hours per square meter are to be used to produce electricity. The solar radiation is comparable with the situation in southern Spain and therefore offers no significant advantage. However, Tunisia's solar plan provides for the implementation of 1,510 megawatts to compensate for its energy deficit. The Moroccan solar plan is thus lagging behind by about ten years.

A prominently discussed project in Tunisia is called TuNur and according to the plans of Nur Energie Ltd. from London it should be developed in the Tunisian Sahara. CSP power plants with a capacity of 2,250 megawatts CSP could be connected to the Italian power grid via a submarine cable and thus supply renewable energies for Europe from North African sources. The expert on the project is Till Stenzel. The young manager supervised the preparations as CEO of TuNur. Stenzel lived in Tunisia for two years to study and talk to the officials.

Tunisia has a special significance for Dii, because it has had its own office there since 2011, initially headed by former Siemens Tunisia boss René Buchler. Buchler had countless, sometimes very tedious conversations with the monopolist STEG. Also, the former Minister of Industry and Development, Abdelaziz Rassâa, was a member of the Advisory Board of Dii. The Tunisian Mouldi Miled represented the state utility STEG for some time in the Dii shareholder meetings, but despite announcements and negotiations STEG never became an official shareholder of Dii. He founded and runs the *Desertec University Network*.

1.4 Egypt

The country on the Nile is one of the geostrategically more important countries in North Africa because of the Suez Canal.

The canal connects the Red Sea with the Mediterranean Sea. Eight percent of all world trade goes through the waterway every day. If the canal were closed, the transport to Europe would be extended by 15 days, to the USA by 10 days – with the corresponding additional costs. Egypt fought five wars with Israel and even the time after the Arabellion was not characterized by stability. After two partly violent changes of government, the elected General Abdul Fatah Al-Sisi holds the office of Egyptian President. At a time when Islamist terrorist attacks are taking place, he is trying to ensure stability and boost the economy of the Arab country with the largest population of around 100 million. In the international media, there is repeated talk of considerable human rights violations and limited freedom of the press. The Egyptians do not like to talk about politics in their country, the risk of attracting the attention of the system through a critical statement is too great – a sign of a military dictatorship. Especially in the western democracies, the question of human rights in this North African country with such a great history is raised again and again. The authoritarian practices of the Egyptian population according to western standards do not encounter any significant resistance among the Egyptian population. One reason is Al-Sisi's legitimization strategies: he justifies the behaviour of his regime with the fight against terrorism and religious extremism.

The Nile is of enormous importance for the country. The hydroelectric power plants at the Aswan dam cover about eight percent of the energy supply. When it was commissioned in 1960, the dam with an installed capacity of around 2.2 gigawatts supplied 75 percent of the country's electricity. The fertile Nile valley has one of the highest population densities in the world. 95 percent of the Egyptians live here. In addition to producing electricity, the Aswan Dam facilitates the irrigation of agricultural land and thus secures the food supply of the population.

Research trip of Thomas Isenburg to the wind farm at the Red Sea

June 2015

In preparation for a trip to the wind farms on the Red Sea in Egypt, Isenburg has been accredited for the state visit of the Egyptian President Al-Sisi in June 2015 in Berlin. Near the Brandenburg Gate the Al-Sisi supporters demonstrate with their national flags. Their president is waving from the window of the Hotel Adlon, there is some cheering. He asks a demonstrator how safe he can feel on a trip to Egypt. His answer: “Travel without fear, now there is peace again.” Shortly afterwards a black limousine drives up, accompanied by police officers on motorcycles. The entrance to the hotel is surrounded by police. Accompanied by bodyguards, Al-Sisi gets into the car and has disappeared again.

In front of the Chancellery a different picture. Here Egyptians from the environment of the Muslim Brotherhood demonstrate. They denounce human rights violations and chant: “Al-Sisi, you’re a murderer!” Some of them pray on prayer rugs. Both groups are separated by security forces. Their countrymen in front of the Brandenburg Gate call them ‘paid Egyptian cheerers’.

Isenburg goes in the direction of the Ministry of Economics. After three security checks, he enters a large hall full of German and Arab businessmen and journalists. The mood is heated. President Al-Sisi enters the hall, together with Federal Economics Minister Sigmar Gabriel, accompanied by Siemens CEO Joe Kaeser. It is crawling with bodyguards. An Egyptian shouts: “Long live Egypt!” The men sign the 8-billion-euro deal, for which the Federal Republic of Germany provides a Hermes guarantee, a state export credit insurance policy that protects against economic and political risks.

Figure 36: Former Federal Minister of Economics Sigmar Gabriel, President of Egypt Abdul Fatah Al-Sisi and Siemens CEO Joe Kaeser. Source: Thomas Isenburg

Al-Sisi's state visit to Germany is controversial, but Minister Gabriel finds positive words: "Egypt is developing well. The bilateral trade volume with Europe amounts to 4.4 billion euros. The introduction of democracy in particular poses major challenges. We offer to talk openly about the topics, Germany stands at Egypt's side. Terror has no place in society. Egypt is home to 90 million people with a 5,000-year history."
Egyptian President Al-Sisi is interested in economic cooperation with Germany. He wants to introduce human rights, democracy and freedom. The politician and the general emphasize the freedom of religion against the background of Egypt's history and the geostrategic position.
Thoughtfully, Isenburg leaves the room – two cultures have collided here, the atmosphere was tense.

September 2015
It is time, Isenburg flies to Cairo and enjoys the Lufthansa flight. In a hotel near Tahrir Square, the science journalist

finds a room and explores the surroundings. On the square Spanish riders remind him of the demonstrations of the past, security forces, police and military are everywhere. And yet life pulsates in the centre of the Egyptian metropolis.
Initially, Isenburg held talks with representatives of KfW Bank, the German Embassy and Siemens AG. For safety reasons, all journeys must be made by taxi...
A technical consultant from Germany advised Isenburg to talk to the Spanish company Gamesa. The manufacturer of wind turbines is very successful in North Africa. Isenburg followed this advice and the Spaniards took care of obtaining permission to visit the Gabal Elzayt wind farm. Isenburg received permission to visit at short notice and booked a flight to the tourist resort of Hurghada. The Gabal Elzayt wind farm is located about 120 kilometres north of Hurghada on the west coast of the Red Sea.
At three o'clock Isenburg leaves his hotel in Cairo by taxi and then flies further south. At the airport he meets Jose Maria Jimeno Pascual, the service manager of the wind farm that has just been built. The men drive there in a large Toyota SUV. Quickly a good conversation develops between them. The forty-year-old electrical engineer comes from Madrid and has been working for Gamesa since 2007. He explains the company's commitment in North Africa with the cultural proximity between Spaniards and Arabs, which goes back to the occupation of Spain by the Moors.
About 100 windmills with a capacity of 200 megawatts are in operation on the wind farm. A total of about 75 people work here. In Pascual's team there are about 10 Spaniards, all others have an Arabic background.
According to Pascual, the culturally specific thing about the North African employees is the religion around which everything revolves. It can be observed everywhere and

intervenes in people's lives. Every day the Muslims spend about one hour in the mosques. They also need five breaks to pray during work. The extremists are a big problem here, explains the service manager. He estimates there is about one in every 100.
Finally, the men reach the site. They stand in an unreal landscape of sand and scree, covered with black lumps of tar. Isenburg meets some members of the team, young dynamic men, in a construction container. The wind is blowing hard. He is greeted with a salam aleikum – peace be upon you. There is a positive atmosphere in the group. The Spaniard wants to motivate his team to perform at their best. They talk about their work and a little about their backgrounds. Then follows a breath-taking ride through the wind farm.

The North African country possesses fossil resources but has been a net importer in recent years. The situation could improve in the coming period with the development of large gas fields off the Mediterranean coast. Rapid population growth of around 2 percent is causing the economy to grow structurally. Over the past ten years, electricity consumption has grown by approximately 8 percent annually. This regularly leads to energy shortages with power outages. First aid was provided in recent years by the construction of a number of gas-fired power plants. There was also a subsidy system for energy sources such as natural gas, petrol and diesel. It was still responsible for 10% of the state budget in 2015. Since then, the reform of subsidies has been a permanent issue. Although initial cuts in subsidies resulted in a significant increase in the price of oil and gas products, they also relieved the Egyptian budget and contributed to Egypt's payment security and foreign exchange reserves. At the same time, Egypt is increasingly focusing on the expansion of renewable energies; by 2035 they are expected to reach 37 percent. In 2018, around 10 percent of electricity will come from renewable energies.

In September 2014, a feed-in tariff of 14.3 dollarcents per kWh was announced, which was followed in November by a large invitation to tender for the pre-qualification for approx. 1,800 MW PV and up to 1 GW wind. The tender met with enormous interest and over 200 companies took part. In January 2015, a long list of prequalified companies was announced, which was to be implemented after a very well-developed process under the leadership of dedicated NREA chief Sobki. In the following months project development companies were founded, the corresponding land in Benban (approx. 60 kilometres north of Aswan) was allocated and initial preparatory work was carried out. Many small and sometimes completely unknown local developers sold their projects or entered into partnerships with experienced companies with the prospect of realising the projects. After a series of preliminary agreements had already been signed in autumn 2015 to finance a large proportion of the projects, the process came to a halt a few months later. For pretextual reasons, insurmountable hurdles were introduced which led the EBRD (European Bank for Reconstruction and Development) and the IFC (private sector arm of the World Bank) to withdraw the funding already promised. This was based on the government's tactic of obtaining a lower price per kWh. In the meantime, less than half of the 14.3 cents/kWh of projects have been awarded in countries such as Jordan and far less in the United Arab Emirates. Ultimately, only one project managed the financial close and successful construction of 50 MW PV in this first phase (IB Vogt from Berlin, financed by Bayerische Landesbank, including Hermes guarantee). What is decisive, however, is that with a delay of over a year, the process for the realization of the Benban complex, known as the world's largest solar park, continued. In the first quarter of 2018, projects with an order of magnitude of approx. 1,500 MW were financed, which have been under construction since the second quarter of 2018. The EBRD and IFC played a decisive role in this, each providing around USD 750 million in debt capital. The loans were syndicated to a large number of international banks with a focus on development banks. In May 2018, Dii was invited to organize a conference in Aswan for the ceremonial opening of the construction works together with the Egyptian Associated Partner Enara. This important milestone in the Egyptian energy industry could be achieved together with the Governor of the province of Aswan, as well as high-ranking representatives from politics and the state utility. In addition, off-grid solutions for oases and remote estates have been realized as government to government projects within the scope of EPC contracts. Masdar from Abu Dhabi, who had already built a 14 MW PV park in Mauritania on the same model, was active here. Like the complex in Ouarzazate, Morocco, the project is intended to develop a lighthouse effect beyond national borders. The removal of the power still poses a challenge. A huge north-south road is to be installed for this purpose. Have her run the power all the way to Cairo.

In addition, Egypt has very good conditions for wind turbines, especially on its coasts on the Red Sea. The west coast is considered one of the best locations in the world. The constant winds are used by numerous wind farms for electricity production, partly financed by the German KfW Bank. The market leader in recent years has been Siemens Gamesa. The Spaniards in particular are particularly successful in intercultural communication with the Muslim North Africans. The wind farms Zafarana 1-8 with 545 MW and Gulf of El-Zayt 2 with 420 MW were installed in 2014.

In 2014, Siemens placed the largest order in its corporate history in Egypt with a volume of 8 billion euros. In addition to several gas-fired power plants with an output of 14 GW, the order also included up to 12 wind farms with up to 600 wind turbines and 2 GW capacity and a factory for rotors in Ain Suchna on the Red Sea. In mid-2018, the Al-Sisi government is striving for rapid economic success. The Egyptians are counting on good cooperation with Germany.

Al-Sisi succeeded in a remarkable piece of energy history in 2015. After some serious power outages in the summer of 2014, an Emergence Power Programme was decided. Within one year, under the leadership of Siemens and in direct contact with Energy Minister Mohamed Shaker, a total power plant capacity of 6700 MW was installed, which solved most of the problems. Many experts had considered the timetable impossible, but in a tremendous effort the plan could actually be implemented successfully and on time. In wind energy, too, there are signs of further massive expansion in 2019. In July, Egyptian Electricity Transmission Company, the Egyptian grid operator, announced the signing of a purchase agreement for electricity from a large wind farm in Ras Ghareb with a total capacity of 500 megawatts (MW) on the Red Sea. The "largest wind farm in Egypt" is being developed by the consortium comprising Orascom Construction Limited (Egypt), ENGIE (France), Toyota Tsusho Corporation/ Eurus Energy Holdings Corporation (Japan). Construction of the park could begin in the second half of 2019. The first privately financed 250 MW wind project of the same consortium in the Gulf of Suez is already under construction. The grid connection here is expected to take place in the second half of 2019.

Egypt relies on microgrids for remote villages and oases. They are intended to replace diesel generators, which were initially responsible for supplying electricity to small towns, with a battery and a photovoltaic system. Individual villages are connected to the system via an electricity grid.

Through its diverse activities, Egypt is developing into a regional centre for the implementation of renewable energies in the electricity grid, alongside Morocco and the United Arab Emirates.

Figure 37: A wind plant in the dessert at the red see

1.5 Jordan

In Jordan, energy costs have traditionally been the highest in the entire MENA region. Jordan could exchange electricity with many countries if the political situation of its neighbours improved. From a geostrategic point of view, the country has one of the most exposed locations in the Middle East, bordering Syria, Iraq, Saudi Arabia, Israel, Palestine and Egypt. That makes it part of the Asian continent. The small Arab country is about as large as Austria and has a population of about 10 million people, just like the alpine country. Due to the many neighbouring states, the last decades have been marked by wars, conflicts and tensions.

Jordan is a barren country with a desert share of 75 percent. The kingdom is the fourth poorest country in the world in per capita water availability. However, there are also hilly areas forested with pine and oak. Similar to Morocco, Jordan does not have significant fossil resources and is therefore dependent on energy imports.

The constitution of 1952 defines a constitutional monarchy of the Hashemite dynasty as the form of government. After that the king has a great power, because he is head of state and commander-in-chief of the armed forces at the same time. The Prime Minister and the Ministerial Council shall be appointed by the King. So far, the royal tribal leaders received the majority in elections. Abdullah II, like Morocco's King Mohammed VI, claims to be descended from the Prophet Mohammed. He enjoys a reputation in the Islamic world as a mediator in various conflicts. At the same time, he is very open to the West and stands by the peace with Israel that his father made in 1994. The country was also flooded with refugees during the Iraq wars and the Syrian conflict, so the population is growing rapidly. The Arab spring hardly brought about a change in monarchical structures. Although unrest also broke out in the capital Aman, unlike in Syria it did not lead to any lasting destabilization of the country.

About 97 percent of Jordan's primary energy needs to be imported. This devours large amounts of capital in the barren and raw material-poor but sunny country. More than 80 percent of energy imports, especially natural gas, were transported from Egypt in 2014 via the Arab Gas Pipeline, which continues on through Syria to Turkey.

Due to the exposed location, the power supply is extremely sensitive to external shocks. This became apparent during the turbulence in 2011, when natural gas transport had to be temporarily interrupted and power generation switched to HFO (Heavy Fuel Oil), which was extremely expensive and harmful to the environment. Jordan is therefore trying to build its energy market on several pillars. The country also relies on Liquefied Natural Gas (LNG) to import fossil resources. This is brought from Qatar by ship over the Red Sea through the Gulf of Aqaba to the Jordanian port of the same name. The first transport from Qatar took place in 2015. Aqaba is the only port city in Jordan to import LNG and export phosphate fertiliser. Jordan, like other states in the region, is also considering the introduction of nuclear power with the assistance of Russia, which is considered unrealistic by most experts. Renewable energies are to play a role in the energy mix of the future because of the climatically favourable conditions. The strategy for this was developed in the heydays of Desertec. Prince Hassan bin Talal of the Royal Family was President of the Club of Rome from 2000 to 2006. According to Jordanian considerations, the costs for the procurement of primary energy should be reduced by 30 percent, thereby reducing the dependence on energy imports from 97 to 65 percent. This also includes the introduction of a 10 percent share of renewable energies into the power plant mix by 2020, by which time a power plant capacity of 600 megawatts of photovoltaic plants and 1,200 megawatts of wind power plants is to be expanded. Currently, development is proceeding relatively according to plan and there are a number of promising projects in Jordan. Success can be attributed to photovoltaics, because in October 2016 the Shams Ma'an Photovoltaic Park with 200 megawatts, currently the largest project in Jordan, was put into operation.

Another promising approach is the Renewable Energy Act, which is extremely progressive for the Arab world. It regulates the feeding of electricity from renewable sources into the electricity grid and has been in force since 2012. Private and international investors are to be attracted for this purpose. There is support from the European Bank for Reconstruction and Development and the Islamic Development Bank amounting to 54 and 5 million US dollars respectively.

Suppliers should propose the lowest possible feed-in tariff. The company with the lowest tariff is awarded the contract. After 14-17 cents in the first round, in the second round, which was published in 2014 and implemented from 2016, the prices for four PV systems of 50 megawatts each were between 6 and 8 dollarcents per kilowatt-hour. Compared to prices of around 20 dollarcents per kilowatt-hour for fossil energy sources, these are competitive in a country poor in raw materials. Furthermore, a further market segment has emerged in recent years due to the progressive legal framework: medium-sized plants ranging from a few hundred kilowatt-hours to a few megawatt hours, which are primarily intended for the own consumption of industrial and commercial customers. Here, companies also take advantage of a special feature, the so-called power wheeling. A PV system can be installed at location A and the electricity consumed at location B. The supplier only has to pay a transit fee to the distribution system operator. In this segment, a sustainable market of several dozen megawatts per year has emerged in recent years; other countries are studying the Jordan success model to see how far it can be transferred to new markets. A pioneer in this market segment is the developer Yellow Door from Dubai.

The third round of tenders for major projects is already underway and the expansion of the 400 kilovolt south-north corridor of the high-voltage line is progressing well. It is a prerequisite for integrating new wind and PV projects into the power grid.

In addition to the tender rounds for PV, various operational wind projects and the important market segment addressed for industrial and commercial customers, there were various EPC projects for universities, administration and major EPC projects such as in Al-Qweira with almost 100 megawatts. To this end, there are bilateral projects with ministries. The breadth of the market in Jordan has also created many local jobs, for example for installers.

The royal family, which has equipped its palace with photovoltaic systems, is also committed to renewable energies. The country's approximately 2,000 mosques are also to be equipped with an energy supply system based on photovoltaic systems – another parallel to Morocco. However, environmental awareness is not yet particularly pronounced in the Arab world.

Jordan is one of the pioneers of renewable energies alongside Morocco and the United Arab Emirates.

1.6 United Arab Emirates

The situation of the Persian Gulf countries differs markedly from that of energy importing countries such as Morocco, Tunisia and Jordan. With the exception of Bahrain, they have fossil resources, which improves the conditions for an energy revolution from an economic point of view.

The area occupied by the United Arab Emirates is only slightly smaller than that of Austria. In total, over 9 million people live there. People from all over the world come to the small state on the Persian Gulf because of its wealth. Architects and businessmen from Western Europe or the USA, stewardesses from Ukraine or Russia, security services from Kenya, maids and hotel employees from the Philippines as well as construction workers from Nepal, India or Pakistan. These hands are indispensable for the gigantic construction projects in the metropolises of the desert state. The immigration is supported by favourable residence regulations: Those who have a job can stay and get a visa. Dubai in particular has an international flair. Numerous skyscrapers give the impression of an Arab New York. The lifestyle is very liberal for the Arab world. It rarely rains in the Arab Emirates, sometimes only two days a year. About two thirds of the country are covered by sandy deserts. The water demand is huge, because of the many artificial green areas and is covered by seawater desalination plants, with enormous (still fossil) energy requirements. In cities such as Dubai and Abu Dhabi, the energy requirement for cooling is particularly important: it accounts for 70 percent of electricity consumption during the summer months.

For a long time, the Persian Gulf region was dominated by England. With the retreat of the British, the fear of attacks of the neighbours Saudi Arabia and Iran grew on the territory of the Gulf State. This caused concern in the then existing seven individual Emirates. In order to defend themselves against their neighbours, the ruling sheikhs founded the United Arab Emirates in 1971.

Huge oil deposits have been identified around the capital Abu Dhabi and Dubai has been an important trading centre for around a 1,000 years. The United Arab Emirates is one of the richest countries in the world. In the Gulf State, the distribution of power is divided equally and the succession to the throne is inherited. There are professionally managed state structures in which representatives of the population are increasingly involved. Dubai has developed into a metropolis of the international business world. The geographical location favours this process, because you can reach Europe, China, India, Russia, Saudi Arabia or South Africa in just a few hours by plane.

The great prosperity does not yet lead to a particularly climate-friendly lifestyle, but on the contrary to large carbon dioxide emissions. That is supposed to change. In January 2017, Sheikh Mohammed bin Rashid Al Maktoum, Prime Minister of the United Arab Emirates and ruler of Dubai, presented his country's new energy strategy. The oil-rich Gulf State wants to invest 153 billion euros in renewable energies. By the year 2050 they are to have a share of 44 percent in the energy mix. This could reduce the horrendous carbon dioxide emissions by 70 percent. The sheiks want to reconcile economic and ecological goals.

However, this is also to be achieved with the help of nuclear power, which is considered clean in the United Arab Emirates. South Korean companies are currently building four nuclear reactors, which will gradually feed a total of around 2.4 gigawatts per year into the grid from 2019. However, the question remains as to whether all nuclear power units will actually still be built in view of the dramatically lower costs for solar energy.

There have already been remarkable successes in the implementation of renewable energies. The 100-megawatt Shams 1 parabolic trough power plant went into operation in 2013. Shams is the Arabic word for sun. Dii shareholder Abengoa is also a member of the operating consortium but had to sell the stake as a result of insolvency. The power plant was one of the largest of its type in the world at the time of commissioning, but does not have thermal storage with liquid salt tanks, instead it has huge gas boosters.

Figure 38: Photovoltaic power plant in the United Arab Emirates. Source: Thomas Isenburg

Another pioneering achievement is the 1,000-megawatt solar park Rashid Al Maktoum. By 2020 it should have reached its full capacity. The first 13 megawatts were already connected to the grid in 2013. After the record price of the second phase of 100 megawatts of just under 6 cents per kilowatt-hour in November 2014 (the contract was awarded to Dii shareholder ACWA Power), the capacity was immediately increased from 100 to 200 megawatts. As a result, the 1,000 megawatts planned for 2030 were already envisaged for 2020, and the target for 2030 was quintupled! A surprising breakthrough for solar energy. For the first time, PV was able to compete with fossil fuels during the day. Further development was rapid: just one year later, the prize for the third round, 3 cents, was announced. This made PV energy the cheapest option and this in such a short time that even the greatest optimists had not expected. As part of these expansion plans for renewable energies, Siemens plans to integrate a hydrogen electrolysis plant, as the company announced in February 2018. The aim is to build the first solar-powered hydrogen electrolysis plant in the region in the open air. This is a pilot project to test how the gas produced can be stored and used for electricity generation or transport.

For only around 2.4 eurocents per kilowatt-hour, a Japanese-Chinese consortium is offering solar power from the 1,200 megawatt Sweihan solar park south of Dubai, breaking another price record in the region and even worldwide in 2016. The large-scale PV plant was built by an Indian EPC contractor, with important parts contributed by the Bavarian associated partner Krinner. This means that renewable energies in the region are competitive or even unrivalled cheap compared to fossil primary raw materials. Further tenders are expected soon for Abu Dhabi.

In addition to major projects, a new market segment has also established itself in the recent past. Similar to Jordan, a so-called net metering scheme, also known as Shams Dubai, was introduced in Dubai in March 2015. Ultimately, the hurdles for industrial and commercial customers are significantly higher, which has led to a relatively slow start. However, after just about 25 megawatts of installed, primarily roof-mounted systems, a strong increase could be observed in 2018, also thanks to a new initiative by DEWA with Etihad Esco. A new installed capacity of 60-80 megawatts is now expected for 2018 alone, with figures set to rise further in the coming years.

The now mature start-up Enerwhere (founded by a German) has introduced an innovative business model in the United Arab Emirates in which large mobile PV fields are used as fuel savers in diesel generators. Over 10 megawatts of projects have now also been realised in this off-grid market segment, and further growth is expected.

Figure 39: Rooft to PV in Dubai Source: Enerwhere DMCC

Another argument for investing in renewable energies in this Gulf country is the low country risk, which for banks is comparable to conditions in Western Europe. This reduces financing costs considerably and facilitates the expansion of renewable energies.

1.7 *Saudi Arabia*

In Saudi Arabia, which occupies about 80 percent of the Arabian Peninsula, two characteristics are particularly striking: the most influential country on the Arabian Peninsula is still one of the most important oil suppliers in the world today. But Saudi Arabia is also an absolute monarchy, which sometimes has medieval traits. Human rights violations are on the agenda. It is characterized by a conservative interpretation of Islam and the consistent implementation of Sharia law. It is especially hard on women.

Crown Prince Mohammed bin Salman, who represents his father King Salman bin Abdulaziz Al Saud in day-to-day government, is expected to make changes towards more liberal structures. This also includes modernising the economy. For a long time, Saudi Arabia's wealth in the form of oil was bubbling out of the desert soil. The necessary momentum in the largest Arab economy was provided by immigrant workers. Eleven million Indians, Pakistanis and Filipinos took over the work in the private sector while the Saudis worked in lucrative state jobs. The dependence on oil billions has created an explosive situation with the onset of the oil price decline. Young Saudis are often unemployed and the economy is uncompetitive. So far, 90 percent of government revenue comes from oil. At the same time there is a youth unemployment rate of 40 percent. The strategy of taking young Saudis into the civil service is no longer working.

The 25 million inhabitants of the Islamic Kingdom have a high energy requirement for seawater desalination plants and for air-conditioning the rooms. Renewable energies should now step into the breach. The lower prices have made them competitive. The world's largest oil exporter wants to restructure its economy with a National Renewable Energy Programme. The budget available for this over the next five years is 30 to 50 billion dollars. This should lead to a 10 percent share of renewable energies in the energy mix. However, there is still a lack of a regulatory framework which, for example, regulates grid access and provides technical specifications for plant operation and contract drafting procedures. That is why the powerful Crown Prince Mohammed bin Salman developed the *Saudi Vision 2030*.

For a long time, the conditions for PV systems were rather difficult, because the costs for fossil raw materials in the oil-rich country are low. Many Saudis see the supply of cheap energy as a fundamental right. It was not until January 2017 that electricity and petrol prices were significantly increased for the first time, albeit from a very low level. Further increases followed and the end of subsidies for fossil energy is now finally in sight. The higher costs for renewable energies could not be justified in this context. Now, however, the situation has changed with the fall in prices for solar energy and wind turbines.

In the spring of 2017, the first tenders for 400 megawatts of wind power and 300 megawatts of photovoltaics were launched. The first concrete actions thus follow the political announcements. In the second round, capacities of 1,020 megawatts will be put out to tender. These are to be allocated to 400 megawatts of wind power and 620 megawatts of photovoltaic power.

The Saudi Arabian authorities in particular have announced major renewable energy projects in the recent past. In October 2017, the Crown Prince announced to his subjects the development of a 500-billion-dollar ultramodern megacity, NEOM, in the northwest of the country. In the spring of 2018, Saudi Arabia trumped off with a gigantic project as part of the restructuring of its economy. Crown Prince Mohammed wants to build two solar fields with a capacity of three and 4.2 gigawatts with the Japanese technology group Softbank by 2019 – that would be by far the largest solar plant in the world by today's standards. The sovereign wealth fund from Saudi Arabia will also be the biggest supporter of the Vision Fund initiated by Softbank with over 40 billion US dollars.

However, such announcements do not ensure that the projects will be implemented. The best example of this is the 54-gigawatt Renewable Energy Programme, which was presented to the public by Khalid Suleiman, then head of the King Abdullah City for Atomic and Renewable Energy, at the third annual conference of Dii in Berlin in autumn 2012. The project was never realised due to a lack of competence between different authorities, ministries and Saudi Aramco. This has led to the frustration of those international developers and manufacturers who had opened local representative offices after the big announcement. Local players such as Saudi Cables and Adecco Solar, who have been loyal Dii supporters over the years, were also very disappointed with the big announcements, which were not followed by concrete steps.

For the near future, the Kingdom formulated key dates. The expansion of renewable energies should reach 9.5 gigawatts by 2023. The lowest prices for PV electricity in the world to date have also been achieved, with some under 2 dollarcents per kilowatt-hour. In the summer of 2018, the Saudis began to take an interest in electric mobility. They joined the American manufacturer of electric vehicles Tesla and seem to want to participate in its competitors as well. Unsurprisingly, one of the key figures in Saudi Arabia is ACWA Power, which , after extremely successful CSP, PV and wind projects, can now finally become active in its home market. In the first tender for 400 MW wind, EDF from France, supported by Masdar, offered the lowest price. An interesting new market is also emerging in Saudi Arabia with the segment for industrial and commercial customers. Dii was able to advise on the implementation of the legal framework. The main obstacle is the still subsidised electricity prices. However, the trend is clearly towards higher electricity prices and a gradual end to subsidies for fossil fuels.

2. Sent to the desert? – An interview by Thomas Isenburg with Paul van Son

Mr. Van Son, a Dutchman at Desertec, how did this come about?

The mainly German shareholders intensively prepared the Desertec Initiative in summer 2009 and were looking for two managing directors with the help of Egon Zehnder. One for the investment strategy and one for the technology. At first, they only looked around in Germany. Although there were many candidates, there was no satisfactory result in the first rounds. Then came the idea of expanding the circle of candidates internationally, because Desertec would eventually be supported internationally. Also, they only wanted one boss. That is how I got in the picture. As a Dutchman, you are as European in the eyes of the Arab Community as any German, but it gives the initiative a more international touch, that was the idea. The *Financial Times Deutschland* said I was sent to the desert. That really did happen in every respect and I liked it very much.

After so many years in Germany, do you feel more like a Dutchman or a German?

I feel like a citizen of the world in the first place. And European, Dutch and half German. In Holland, friends catch me more and more often with Germanisms in my language and in Germany after two words they immediately hear that I am Dutch. In the past, people used to rave about Dutch tolerance, the cute *kopje koffie*, Johan Cruijff and Rudi Carrell. Today, we come to the topic of Geert Wilders with his populism and the absence at the World Cup in Russia. Holland is no longer a *Frau Antje* fairy tale in Germany or 'windmills, wooden shoes and tulips', like the old image internationally. Many countries are currently struggling with the same issues and there are few pronounced differences. The relationship between the Netherlands and Germany has developed very positively in several generations. They are exemplary neighbours and they are both committed to a strong Europe and a pluralistic and open world. When I think about my parents, I know how different it used to be. Their story is shocking, but it has brought me great curiosity and fascination for Germany.

What has shaped you so much?

My father, Theo van Son, was 20 years old and the son of a blacksmith when the Wehrmacht invaded the Netherlands. He lived in Loosduinen, a village near The Hague. In Rotterdam, about 20 kilometres away, bombs fell. Instead of staying home safe, he was driven by youthful curiosity. With some friends, they took their bicycles and cycled towards the big fire. Shortly before reaching his destination, he was hit by flying fragments of shell on the outskirts of the city. He was seriously injured and was sent to one of the military hospitals. He miraculously survived but had to give up his profession as a blacksmith. He did not allow himself to be defeated and after the war became an inspiring Secretary General of the Dutch Farmers' Union and later organizer of the immigration of workers from the Mediterranean countries. He encouraged me to work with people of any religion rase or country for what we today call `sustainability. He gave me some simple, practical wisdom from the peasant world that came to mind when I wrote this book: *Ontdekken, Dekken, Doen* (discovering, thinking, doing). According to him, the people with whom one has to deal in life can be divided into: *Koplopers, Meelopers en Klaplopers* (pioneers, opportunists and freeloaders). Throughout the process of the Desertec Initiative I experienced the amazing power of these simple insights.

What were your first professional experiences?

In November 1979, after studying power engineering in Delft and completing my military service, I started my first job at Siemens in Erlangen. The development of mathematical programmes for power plant and grid management was my task. I was curious about my father's picture of Germany. How would my emigration (at that time I still felt that way) to Germany be perceived by him? But my father was proud of me and he showed no hate or resentment. He said to me: "Hats off for the land and the peaceful power that we are now experiencing. You can learn something there. Show me what you can do." This was followed by turbulent years at the forefront of new developments, at Siemens in Germany, then at TenneT's forerunner SEP in the Netherlands, at the US consulting company Energy&Control Consultants, later taken over by DHV/KEMA, at Essent, the Dutch market leader in (green and traditional) energy supply, as head of their German branch , and briefly at Econcern, a fast-growing company for renewable energies, which painfully went under during the financial crisis in mid-2009.

So how did the Apollo project Dii get started?

On 31 October 2009 I was appointed Managing Director of the Desertec Industrial Initiative, Dii GmbH. Who would have ever dreamed that a Dutchman would get the best job of the year, as a German DAX board member called it? I am proud, proud of the Desertec movement, proud of my father, who died in 1999, and proud of Germany, which here became the driving force behind a fascinating idea. It was a historic moment for Gerhard Knies, the industry, the international energy revolution, the international community and, last but not least, for me.

How do you see the start now, looking back almost 10 years later?

Desertec's announcement made a big splash in Germany and even worldwide in 2009. Although Desertec's main objectives for 2050 were much broader, the media focused from the outset mainly on the intention to transport electricity from the deserts to Germany and Europe at relatively short notice, thus replacing nuclear energy and fossil-fuel power plants in Europe and reducing dependence on Russian gas. The industrial movement, which emerged from Desertec under the name Dii or Desertec Industrial Initiative, was also strongly influenced by the interests of the solar thermal industry. They had no growth prospects in Germany. North Africa, on the other hand, was still completely new territory, with attractive prospects for solar thermal technology. According to the intentions of those stakeholders, Germany should promote such power plants and build grid connections in the Mediterranean. Desertec was thus misunderstood by many as a one-dimensional major project for Europe.

A misleading starting position does not necessarily mean the end of a realistic, socially acceptable and future-oriented movement. The changeability and assertiveness of Dii has clearly shown this. A team of highly motivated talents, at the start of their professional careers, supported by many companies, institutions and governments, have enthusiastically accomplished almost impossible things in less than nine years. It has been instrumental in convincing the renewable energy market in North Africa and the Middle East to move forward. Everyone who has contributed to this can be proud of it. This applies in particular to the tough comrades-in-arms in those intensive years. They're my best friends.

How did the trial go?

Through system studies, socio-economic analyses, many discussions with stakeholders in Europe, North Africa, West Asia and China as well as hot discussions, accompanied by conflicts of interest, we were able to separate the sense from the nonsense of the desert power. The road to a local and international market for desert energy was bumpy. Above all, it was necessary to settle accounts with incorrect, fixed ideas and opinions and to further develop the emission-free perspectives for the coming decades. In the first place `local ownship should be ensured. I had the pleasure of personally travelling to almost all of MENA's desert countries and talking to decisionmakers and experts there. The programme also included trips to the vast and very diverse deserts and contacts with people from civil society. That opened my eyes. We came to the conclusion that the key lies not with idealistic groups in Europe or in the fight against fossil top dogs, but with governments and companies that really understand the economic perspectives. We found the smartest and most effective pioneers in Morocco and the Emirates. Decisionmakers in the region are looking to the pioneers. Good role models will easily find followers, whether out of jealousy or because of other motives.

We had to deal with many questions and concerns. Let me give you a few examples. It is quite understandable that in highly industrialised, fossilised Germany, the question arises of how to ensure the supply of electricity in the long term without fossil emissions. No company likes to be confronted with the phenomenon of stranded investments. On the other hand, the fossil age is irrevocably coming to an end. The good news is that, in principle, Europe's renewable sources are unlimited in their ability to cover European consumption. An exchange with the big neighbour, MENA, is therefore not necessary per se, but in the long run it will greatly reduce energy costs. Russian gas and other carbon carriers are in the best case only a bridge energy. There are also many concerns as to whether connections between the major renewable sources and the consumption centres via electricity grids or other transport routes – for example via hydrogen or alternative synthetic fuels – are safe. Our studies and practical examples show that all this can be economically guaranteed. Energy exchange will lead to a better, greener energy mix in the associated markets and thus reduce emissions. The elegant thing is that it doesn't matter where emissions are reduced if only it is effectively on balance!

The issue of expanding the electricity grid is being observed with great scepticism, particularly in Germany. It is true that networks in densely populated regions can hardly be expanded sensibly. There is a large interconnected system, for example

Europe and MENA, but there is a range of alternative solutions. The cross-border interaction of electricity producers (soon also hydrogen producers), network operators, energy trading, distribution, flexible consumption and storage is often still limited by national thinking. But cooperation, coordination and exchange are immediately worthwhile. Dii has always positions itself as a signpost in these questions.

Where is Dii 2019?

There is certainly no all-encompassing miracle solution for the development of desert energy. All roads lead to Rome, so to speak. There are as many opinions on the solutions in each region as there are players. Now that the energy revolution in MENA is underway, Dii only has a modest role to play in guiding the industry. It cooperates with the many key players in the region. As far as Dii is concerned, the main thing is that, through international cooperation, desert countries will be able to develop their renewable sources competitively, build infrastructure, reduce fossil fuel production and create new jobs. So, don't generate unnecessary subsidies but ensure strict measures to reduce emissions. Renewables have been competitive since about 2015, and in a healthy market environment they will assert themselves in a relaxed manner. We belive that a strong and effective penalty on harmful emissions and based on on free market mechanism MENA will become, so to say, automatically a power house to its own citizens and industry and to the world. That is why we mainly deal with practical questions to accelerate the process. Our target direction is summarized as core: Our Mission: No Emissions.

How was Dii accepted?

As described in our book, Dii has never lacked the attention of German and international celebrities from politics and industry. In 2009 it stood for great media interest. European parliamentarians, institutes, politicians, universities, NGOs, in Germany and worldwide, reacted with great curiosity, scepticism or high expectations to the announcements of the companies. Federal Chancellor Angela Merkel already welcomed the Industry Initiative Dii in the summer of 2009. Hermann Scheer, whom I spoke to shortly before his death, was perceived as an inconvenient critic of Desertec and Dii, even though I found his criticisms in themselves easy to digest. Internationally there have been laudatory, critical and even outraged statements. President Sarkozy feared a dominance of Germany in his backyard the Maghreb. He would soon start a similar initiative: Transgreen, later renamed MEDGRID. His advisor, Christian Stoffaës, visited me immediately after my employment and introduced me to the French-German subtleties. The acceptance in the MENA countries was, as described in this book, very different and ranged from greatly exaggerated expectations to total disinterest. We listened patiently and tried to make it clear that Desertec is not an army invading MENA, but the beginning of a unique intercontinental collaboration with an open exit.

Klaus Töpfer, the former Environment Minister and UN Envoy, was my advisor for some time. He was disappointed that the industry, on the one hand, made big announcements but, at the same time, showed no hard commitment. I told him it was way too early for that. In the beginning, the Desertec concept was not much more than some groundbreaking but still very rudimentary studies. Talks with ministers, secretaries of state and parliamentarians also followed with Klaus Töpfer. One could not expect that everyone would see through all the complexity, but his deep knowledge of the players and his extraordinary eloquence opened many doors for us. The sympathy we gained for our studies helped us a lot afterwards.

Has Germany taken its chance?

To my astonishment, German industry did not make the buest use of its good starting position in MENA, or used it only to a limited extent. German companies are generally highly rated in MENA. In the energy industry in MENA there are traditionally successful German manufacturers and suppliers of hardware such as Siemens, ABB, Leonie or service providers and engineering offices such as ILF, (formely) Lahmeyer and Fichtner. For example, Siemens was and is successfully present in the region with gas turbines and grid systems. It has continued to drive the wind business in MENA but hardly plays any role in the solar sector. German companies are generally very cautious in MENA. Investment protection is required, but they are often reluctant to take the trouble to build up the necessary long-term partnerships and friendships in the region. Apart from these categories, German project developers currently (2018) do not play a key role in energy transition in the region. The German political flanking was very committed and effective at the beginning, but lately it has become unconvincing. However, there are refreshing GIZ and AHK initiatives here and there on practical energy transition issues. Germany deserves praise for the great Desertec initiative, but in my opinion, it is not taking advantage of the good starting position. Fortunately, a number of local companies in the region, such as ACWA Power, Access Power, Masdar, Fotowatio, Abdul Latif Jameel, MASEN, Nareva, Orascom NEOM and others are now doing this. But French , Chinese, Italian, Japanese and South Korean developers are also increasingly taking advantage of their opportunities. The question is what the second round of energy transition in MENA will look like. Since immeasurable amounts of wind and solar energy will soon be available in MENA for a few eurocents per kilowatt-hour and stored and transported in various ways, this region may become the world's largest energy market. This should also sound like an Eldorado for the energy-intensive industry. German industry will have many new opportunities, but will it have the courage and patience to use them in partnership with local players?

Did Desertec fail?

We have hear that a lot, especially in Germany. Unfortunately, journalists often don't know any better. How can a long-term movement for emission-free energy from the deserts in 2050 have failed after only a few years, even though we can see everywhere that developments are rapidly asserting themselves? The fact that at the moment neither Germany nor Europe have short or medium-term power generation bottlenecks makes the allegations of failure even funny. In reality, everything is much more relaxed and natural than we thought at the beginning. In the German media it is easy to be considered a failure. Every day something or someone fails theatrically in the media. Can one also learn something from initial problems and turbulences? If something about Desertec has failed, then perhaps a constructive and clever discourse, a fair, open debate in a search process for all. Germany does not lack money, organisational power, industrial culture, good and above all great ideas, good people, good companies or a capable government and a stable democracy in itself. However, the way in which stakeholders in energy transition learn from new knowledge and improve is not yet worthy of praise. Biased points of view are often contested with a lot of force, but clever new solutions or anticipating new market conditions are often difficult to negotiate. It is to be hoped that the new generation of thinkers and developers will further strengthen the signs of a more open culture emerging in recent years. The energy transition is taking place worldwide. Germany used to be 'first mover'. That is over now. But Germany can take advantage of new opportunities in the area of *Power to X*, for example, if it gets its act together in time.

What is your conclusion?

Desertec, or better phrased sustainable energy from desert regions, will undoubtedly be realised, but first and foremost driven by the countries in the region for their own needs. The export of oil and gas from the region will not disappear immediately, but an alternative in the form of desert electricity, gases or liquids will emerge at lightning speed. Desert energy will assert itself faster than one can imagine today. Smart international companies anticipate this and form local partnerships at an early stage. They courageously seize their opportunities in this politically sensitive region. Companies that can count on strong support from their own government have a strong advantage. As Gerhard Knies announced at the time, the deserts will become the great emission-free energy suppliers of the future. I hope to live to see it.

Figure 40: Heads of the movement: "Emission-free energy from the deserts". From the left to the right: Ahmad Ali Al Ebrahim, Martin Herrmann, Frank Wouters, Khalid Rashid Al Zayani, Paddy Padmanathan, Paul van Son, Ad van Wijk. Source: Dii.

A day in the life of desert power traveller Paul van Son

Recorded by Gerhard Hofmann, journalist and consultant of Dii

20 October 2011, 8.26 a.m. – Berlin Tegel, gate A 03. The most practical airport in the world will soon be replaced by Schoenefeld at the gates of the capital – then the paths will become longer. Flight

AB6192 from Munich is on time – Paul van Son even lands four minutes earlier than planned. Three and a half hours before, at five o'clock, he got up in Großpienzenau in the Bavarian Oberland, then drove an hour by car to the Franz-Josef-Strauß-Flughafen. The flight was fully booked – all business travellers. But Van Son is lucky: although he only got an unloved middle seat, the seat next to him remains empty – at least some space for reading the newspaper ("Thank God the pages are getting smaller and smaller."), when the legroom is getting tighter and tighter: "These flights are more and more becoming like a sardine transport", he states with his typical dry humour. The Dutchman can adapt to any situation – the story about the frog that falls into the cream and then struggles until he sits on a lump of butter the next day, and in retrospect regards this as an interesting experience, probably applies to him. There is no breakfast anymore on these flights; but that it not necessarily a bad thing – years ago a stewardess replied to the question of "whether you could eat that" with "rather not".

In Berlin, a full day awaits him – meeting after meeting. The occasion is the Energy Forum of Ghorfa, the German-Arab Chamber of Commerce and Industry, in the Nobelhotel Adlon at the invitation of Secretary General Abdulaziz Al Mikhalfi.

First reading his mail in a taxi – Paul van Son always takes his iPhone to the limits of its capabilities, always on the lookout for charging options. However, you rarely have to wait long for a reply to an e-mail or text message.

Today the timing is indeed tight: two meetings with members of the Bundestag, the preparation for a meeting with Jochen Homann, State Secretary in the Federal Ministry of Economics, the conference itself, in between a podium in the working group 'Education and Know-How Transfer along the Energy Value Chain – Manufacturing, Project Development and Operation' during the Energy Forum; in between lots of contacts, greetings, reunions. In the taxi, the first appointments are made, the daily routine is structured, in short, it is considered to postpone one of the MdB appointments, then the decision that Dii would join the Future Energies Forum, an industry-neutral and politically independent discussion forum which , according to its own statement, organises "the energy policy dialogue in the run-up to parliamentary decisions" in Berlin and Brussels. Finally, a brief outlook on the second annual conference of Dii on 2 and 3 November in Cairo. In addition to Paul van Son's lectures, such conferences have two decisive aspects: "The media echo when you announce the conference, and the breaks for networking."

When Paul van Son enters the conference area of the Adlon, the forum visitors already stand shoulder to shoulder. Shaking hands, here a quick word of recognition, there a fleeting nod, some are more familiar than others – people know each other in the energy community, especially the renewable one. Jürgen Hogrefe, Ghorfa board member and Berlin representative of Solar Millennium AG, is very satisfied: "Over 300 participants, half of them from the Arab world – that's very good". Last year he had given the impetus for the first conference of this kind. The German ambassador in Kuwait, Frank Marcus Mann, reports that a rethink is also beginning in the oil-abundant country on the Gulf: From now on, 15 percent of new buildings would have to be supplied with solar energy; of course this will take time, as for the time being only the private sector would be obliged to do so. "Energy is so cheap!"

Handshakes with the ambassadors of Algeria and Morocco, the new man from Tunisia, here the Arabellion provided for the first change in the diplomatic corps. Paul van Son has become a well-known and sought-after man in less than two years.

Ex-Florett World Champion Dr. Thomas Bach ('doctor juris utriusque' it says in his CV; the World Championship title is not mentioned) welcomes everyone in his capacity as President of Ghorfa. What does Ghorfa actually mean? is often asked, and only a few know it: Ghorfa is a term from Berber architecture, designates a granary, often several storeys high. "Energy is the backbone of economic growth and prosperity," says Bach, mentioning that the volume of trade with the Arab world has tripled in the past ten years. And he is the first to provide an English translation for the German word *Energiewende*: 'Energy Turnaround' – let's see whether the turnaround will prevail or whether 'Energiewende' will become a loanword in English. Paul van Son sits in the second row in the reserved seats, and he can hardly believe his ears: No conference has ever heard the word Desertec or Dii as often as here and now in the Adlon. All pf the speakers mention the desert power vision or the initiative.

Economic State Secretary Homann is taking it to the top: he dedicates a quarter of his keynote speech to praising the Desertec vision that it could contribute to future development in the Middle East and North Africa and would bring the Arab world closer together through a cross-border energy market. He then announced that he would meet with Dii shareholders and Paul van Son ("Here he is!") at the ministry in the afternoon to discuss how politics, i.e. the federal government, could contribute to the acceleration of Desertec. Paul van Son would know how: Not only by promising billions to save the banks, but also by putting a tiny part of these bank rescue euros into future energies. The State Secretary is raising expectations – but he will not become as concrete in the afternoon, although his speech in the morning gives cause for hope.

Jamila Matar, head of the energy department of the Arab League, lists the partly ambitious plans for the introduction of renewable energies in her region: Morocco 42% by 2020, Algeria 20%. The growing interconnectedness is also breaking down other barriers within the Arab world.

In the break, a short, rather pitiful conversation with the Syrian ambassador: One cannot get beyond politeness ("How do you find the conference?") – no wonder in view of the bad news from his country. It is different with his Iraqi colleague: He even speaks Dutch , Van Son is pleased: He can speak once again in his mother tongue, which is rare. There is an exchange about the chances of CSP ("less"), PV ("quite good") and wind ("some excellent areas in Iraq").

In between, a short question-and-answer game with a very knowledgeable *Handelsblatt* editor: When will the first electricity flow? ("Soon, as early as 2013, when everyone will be involved in politics and business. Homann has just promised support."). What is the point of Dii considering the United Arab Emirates as the location for the next reference project? ("Nothing.") Could Dii's Greek Helios project dig up necessary funding? ("Don't worry about it.")

11.30 a.m. – Waiting in the noble foyer of the Adlon for the first meeting of the members of parliament. People are present here whose circles Paul van Son was not automatically born into. He comes from a Catholic, but Calvinistic family with nine children. As a child he went to school in short trousers and long socks. At that time, life in Holland was still very restricted: The Church had seen with great concern how the peasants had moved to the cities and lost their old values there – one of his father's areas of responsibility at the time was this problem. Paul van Son still remembers how the Dutch society consisted of individual pillars, especially the Catholic and the Calvinist – Catholic majority in the south , Calvinist north of Meuse and Rhine. It was not until the 1960s that the isolation and rejection gradually eased.

11:45. Johannes Vogel, 29, a member of the Bundestag of the FDP since 2009, enters the Adlon foyer fast, relaxed and unconcerned. The young political scientist approaches Van Son, who is almost twice his age, openly and extends his hand to him. Vogel, who was actually the labour market policy spokesman for the FDP parliamentary group (in this respect the successor to Minister Niebel), had described Desertec as the FDP's flagship project at the time. That had met with interest – now he is sitting opposite the Dii boss. He started out politically with the Greens, but then went on to pursue a lightning career with the Liberals. He quickly gets to the point, wants to help ensure that despite the euro crisis Desertec and Dii remain on the agenda, "at least in 10th place"). He wants to talk to the responsible ministers, Rösler and Niebel, whom he knows well, and also inspire them with the desert power idea "which can carry the whole of Europe". The Ministry of Economics and the Ministry of the Environment would have to get on board, the BMZ (Federal Ministry for Economic Cooperation and Development) would have to end its hesitation. "The entire German foreign trade policy must be readjusted," says Vogel, "because the hopes of the people there must not be disappointed." He seems very convincing in his momentum.

1 p.m. – CDU member Hartwig Fischer, who has been a member of the Bundestag since 2002 and a full member of the Committee for Economic Cooperation and Development, acts more deliberately. He is nicknamed 'Afrika-Fischer' in the parliamentary group, because since 2010 he is president of the German Afrika Stiftung e.V., and chairman of the working group for Africa of the CDU/CSU parliamentary group, rapporteur for East and Central Africa and member of the parliamentary groups for Africa. Fischer has worked his way up from simple circumstances, made a retail apprenticeship after finishing elementary school, and today he is a great figure in the Union faction. He's not a man of many words, he wants to know right away what he can do. Fischer was a nuclear proponent, but the political pragmatist is quickly getting ready to leave ("We won't discuss this anymore!") The old prejudice "that we will leave and then import nuclear power from France" drives him on. He knows that it depends on the

production, that the renewables must gradually displace the fossils, that as long as a mixture of everything comes out of the socket. Paul van Son tirelessly presents the Desertec vision (one might ask when this will finally come to an end, but his enthusiasm is still contagious), describes the current situation, the reference project in Morocco, which is now also supported by the King, the efforts to create a European-African market, the hopes and expectations for the meeting with the Economic State Secretary in the afternoon. Fischer wants to play a role in a long-term partnership programme between Europe and Africa, saying that Desertec is not a 'Growian' after all (54 million deutschmarks flowed into the first large wind turbine in 1983, but it turned out to be a flop, although much was learned from it). He warns against too full-bodied announcements, and he is probably right.

1:45. Hurrying back from the parliamentarian restaurant of the Reichstag to the Adlon: Session 3 (brought forward because of Paul van Sons schedule) awaits: 'Education and Know-How Transfer along the Energy Value Chain' – it is about one of the most important aspects besides power generation: If the target countries lack one thing in addition to the development of industry and agriculture, it is education and training. Paul van Son condenses this into the formula 'Bringing the idea to the people' and brings it to a short triad in a short speech from the podium: 'Acceptance – Engagement – Participation'.

At the time, when the 'big plan' with the 400 billion had been announced, politicians in some countries had said: "How? The Germans will be back?" That was "of course a mistake," says Van Son, and that is why Dii had flown out right from the start and travelled to all the countries in question to talk to the sceptical politicians. In the meantime, the Desertec University Network has also contributed its share – through university partnerships, networking of research centres, linking of laboratories, exchange of students, placement of scholarships. In the short round of questions, a participant of Paul van Son would like to know whether a survey on public acceptance has ever been carried out. At the population level, the Dii chief must deny this – "but at the political level, we did it personally".

4 p.m.: Van Son, a traveling salesman in the field of desert power, is already on the move again – now to the E.ON representation in Berlin. There he wants to prepare himself with the other discussion participants from the circle of shareholders for the State Secretary's meeting. After all, there is something at stake: finally, politics should move and pledge its support – and not just in words. And he and his colleagues expect the Federal Ministry of Economics to play a key role. Paul van Son and the others want to introduce Jochen Homann to the 150 MW CSP pilot project, a recent spin-off from the first reference project, and discuss state funding. The fact that Homann was once head of the BMWi's basic foreign trade policy department could be helpful, as could the fact that he had been involved with Desertec on several occasions, such as when Prof. Hans Müller Steinhagen, once father of Desertec studies at the German Aerospace Center, presented the desert power vision to him, or when he represented the German government at the first Dii Annual Conference in Barcelona.

At 5.45 p.m. it is time. It is not easy to get into a Federal Ministry recently, since 9/11. Even an appointment with a secretary of state or minister means first of all waiting, even for one or more CEOs. But the gentlemen don't have to wait long in the reception room at entrance 2 in Scharnhorststraße. A friendly employee guides them to a meeting room on the second floor. State Secretary Jochen Homann receives the five-member Dii delegation cheerfully. There is water, coffee and biscuits. Homann is, as always, perfectly dressed. The former speech writer of the FDP Ministers Hausmann and Bangemann is well prepared by his people, has written documents, four of his co-workers, specialists in the field, are present.

Homann wants to know more about the Dii pilot project in Morocco ("I've waited a long time for this day"), he makes it clear that the Ministry of Economy is willing to help make the project a success. But it has to be "as cost-efficient as possible," he says. From the outset, the need for subsidies must be made credible and degressive – no permanent subsidy pit must be created. Dii and its shareholders would have to take risks themselves if they expected financial participation – without security through state guarantees, at most perhaps through foreign trade promotion funds. The pilot project would not appear as a purely German-Moroccan project, but would be flanked by a European approach. EU Commissioner Oettinger would be involved and important representatives from the various countries would be invited for talks. Homann asks for a to-do list of critical points for the preparation of such a discussion at EU level. However, he does not make binding promises. The aim was to be able to make a clear yes-no decision for the pilot project by the end of the year. After all, the government task force would continue to work on it.

In Paul van Sons' absence, a shareholder talks about jobs created by renewable energies: For Germany alone, Jürgen Wild of the M+W Group estimates 330,000 in science and research, production and the supply chain – the tertiary sectors have not yet been calculated. M+W is currently announcing that it has now reached 7,000 employees – a new record.

Wild: "Global thinking must not ignore the desert regions of this world, which after all cover about one fifth of the land area. That is why we have been working on this topic since the middle of the last decade. We have participated in the first solar thermal test power plants and supported Dii. The first concrete results are now emerging. Construction of a plant in Morocco could begin as early as 2012. The technology – photovoltaic or mirror – is selected primarily on the basis of economic criteria and grid requirements. Solar power plants can fall back on proven technologies, are quickly realisable and environmentally friendly and can support a sustainable economic upswing in the affected regions. I am therefore confident that the share of electricity from the desert in world energy production will grow in the future."

At 19.30, a relatively satisfied, cheerful Dii boss winds his way through the revolving door of the Adlon. Paul van Son has a gift: he can see the positive in everything – even in dark moments he remains calm, serene. In Cologne it is called "Et hot noch immer jot jejange." He keeps shaking new hands. Patiently he gets involved in small talk, answers the same question for the umpteenth time, as for instance: When does the first desert electricity flow into Europe? Van Son is lucky: he sits at the 'Hamburg' table with relaxed guests: the German ambassador in Kuwait, field staff, businesspeople, Arab diplomats. In between a short conversation with Thomas' predecessor Bernd Pfaffenbach , who for many years under three chancellors as a so-called Sherpa helped prepare all summits of the federal government, then a short talk with the SPD faction leader and former Foreign Minister Frank-Walter Steinmeier – it was agreed to talk in detail shortly.

Previously, Steinmeier had held his dinner speech in precise English , as if he were still Foreign Minister. There is great approval on all sides when he does not spare critical remarks addressed to Israel's. Not militarily, only politically the peace in the Near East is attainable. "I myself have often warned my Israeli friends: Don't rely on the status quo! It is not a guarantee of safety!" But he also warns those present: "Israel's isolation cannot be in anyone's interest, not even the Arab world". Then, in keeping with the theme of the evening, he comes to speak on the energy question. Fossil energies would decrease – the question was what we could do to jointly secure energy supplies in the second half of this century. "Not only do we extend a warm welcome. We need you!" He mentions Desertec only afterwards in the interview – positively: "Desertec is already working, because it brings the MENA countries together, even if no electricity should flow to Europe at all for a long time yet."

Six days later, October 26, 9 a.m.: In the Paul-Löbe-Haus (abbreviated to PLH by insiders), all meeting rooms are circular. In E-800, the Committee on Economic Cooperation and Development meets regularly (most committees are named after the ministries assigned to them). Dagmar Wöhrl, the Chairwoman, was Parliamentary State Secretary in the Federal Ministry of Economics during the last legislative period and as such one of the political godmothers of Dii: at the time, at the legendary press conference in Munich Re, she was one of the representatives of the Federal Government.

Wöhrl opens routinely, welcomes those present and congratulates the FDP delegate Günther on his birthday. The media bench on the first floor of E-800 is sparsely occupied, today the interest is focused on another event. Shortly after 12 p.m. Chancellor Merkel will make her government statement on the European rescue plan ("the biggest test of economic and monetary union ever", she will say and "the risk is justifiable"), to one trillion euros, some even claim two, the 440 billion now in the pot is to be raised, a good half of it German euros. Because of the euro crisis, the meeting of the Committee on Economic Cooperation and Development has been brought forward to the early morning. Incidentally, the Chancellor will be the first in this office to visit the Committee on 1 December to underline the importance of this body.

Because of the surprising advance, Paul van Son had to leave his Upper Bavarian residence very early. He flew from Munich to Berlin this morning and is the first of the invited listeners to be asked to speak.

"Dii is a motor for sustainable development," he explains to the nearly two dozen members of parliament sitting in a circle with a somewhat harsh-sounding voice at first, but this quickly subsides – and he gains confidence. It was important for a market to emerge, for renewable energies to become competitive. Again and again, he has explained everywhere that Desertec is not a 'real project', but a multitude of power plant projects of different sizes – including here.

He talks about the most concrete project to date, the 150 MW pilot project, part of the first reference project cluster in the Moroccan desert. Because he knows that in the ranks of SPD and Greens there is still discussion about 'central vs. decentralized', he anticipates this point right away (later one of the members of parliament will say that the Rhine-Main area or Stuttgart cannot be supplied with small decentralized units). The countries of North Africa could replace their gas-oil energy supply with renewable energies and partly export their raw materials.

It is a partnership, the socio-economic aspects should not be neglected as there are: Environmental pollution, reduction of greenhouse gases, water pollution, reduction of water consumption, but above all the prospects for young people. Van Son explains that CSP is more labour-intensive than photovoltaics, for example, and therefore creates more (direct and indirect) jobs. The aim is to develop their own industries in the target countries, to stimulate technology and know-how transfer, to accelerate network expansion, storage technologies and the expansion of the entire infrastructure. The area of research and development is particularly important: the recently founded Desertec University Network is very helpful, for example in the exchange of students and scientists, in knowledge transfer and capacity building, and in making new contacts. This precedes the cooperation of entire industries. He was sure that this would lead to a major energy turnaround between Germany, the EU, the Middle East and North Africa. And Desertec has the effect that the energy turnaround arrives there. As a Dutchman in the German parliament he can say: "The whole world looks to Germany." It is nothing less than the export of the German energy system transformation. Some parliamentarians in the circle, not a few of which until recently were still nuclear power proponents, nod approvingly.

Adel Khalil of the Regional Centre for Renewable Energy and Energy Efficiency in Cairo (RCREEE) speaks of a win-win situation. One square kilometre of desert offers the equivalent of two million barrels of oil. Construction and maintenance of solar power plants would create between three and ten jobs per megawatt output, says Khalil. However, North Africa could not meet the 'huge investment needs' alone.

Albrecht Kaupp, also from Cairo, a widely travelled GIZ energy expert who works from Morocco to Turkey, supplements Van Son excellently: He carefully deals with the questions that the previous speaker dealt with rather cursorily. He says the electricity and gas market must be integrated. The question sent in advance, whether Dii is a development project? Answer: "Yes."

Since Dii has been in the media, one notices increased demand, the quality of the discussion has improved, the polarization is good. "The 400 billion that were put into the room at that time will be distributed over 40 years, so it is not such a big sum."

What current challenges does he see? "In the technical area: only two positions are still open for the electricity ring of the 24 countries around the Mediterranean. Connections by submarine cable between Morocco, Algeria, Tunisia and Libya with Europe are not a problem – the rest would have to go via Turkey to Europe. A pilot plant is needed."

What are the conditions under which electricity can be supplied? "Our tariffs are so low (in the Maghreb) – can we afford the desert current without overthrowing the government? Bread and energy prices are important. Photovoltaics is cheaper... But China will have more of it than we do; CSP is different; Germany, the USA, and England are still the leaders. Water consumption at CSP is no problem thanks to the air cooling."

The future lies certainly in the attempt to create export markets; an important challenge here is to shift 'bad subsidies' (for oil and gas) into 'good subsidies' – into renewable energies: "One cannot and should not abolish the subsidies at all, but shift them, because the current five to eight percent increase in electricity consumption in the region still swallows all increases in renewable energy production anyway."

Green MP Ute Koczy turns to Paul van Son: "Mr. Desertec...!" That's it... everyone's laughing. Would other countries be prepared to approach Morocco? Which and how many jobs would be created? German development aid should set clear signals. How to support that? What other possibilities are there for exporting renewable electricity? Others see "time running out – we need feed-in regulations, and purchase guarantees!" Is there any delay due to the democratic movements? If submarine cables were not a problem, what would happen on the mainland? One question deals with water treatment using photovoltaics. Niema Movasset of the Left finds it "hardly credible that the electricity should first benefit the local people", terrorism threatens, cables and power plants become target of attacks, who then provides security? Would these jobs come from local security companies? SPD member of parliament Sascha Raabe asks about land use, what property rights there are, how is the land left to whom?

Khalil explains that the standard for desalination is residual heat. Property rights to know-how are important; cost reduction is the key criterion: “We would certainly have stability in two years, all governments would make promises and guarantee security.” He also believes that subsidies must be redirected, renewable energies for gas exports or for the chemical industry; land use is not a problem in his eyes, the desert is not used.

Kaupp once again addresses the prices for renewable energies: in the NA countries one has double the irradiation, one can produce for half the price. “Electricity arrives 30% cheaper in the middle of Europe, so it would be competitive.” The price structure in the countries of North Africa is simply a catastrophe; in Egypt on average 2.5 cents for kWh, in other countries 3.7 cents... 1 kWh from diesel costs 17 cents in Morocco, wind only 7 cents, solar 17 cents as well. Water is more important; we have to produce more water than we supply electricity.” Then he warns against the strong nuclear power lobby: it is well on its way there, he says, and is pulling members of parliament and ministers. Decentralised energy supply creates more jobs than centralised energy supply: “There are even two PV production plants, one in Syria, one in Jordan – unfortunately they are not yet competitive.”

Paul van Son concludes: decentralised – centralised – both are important. Dii wants to stimulate larger plants: around 500 MW. He reports concrete news from Morocco. “With MASEN, a first pilot of 150 MW (from the cluster of 500 MW) is decoupled. Our partners have already been asked whether they can provide equity. Yeah, they said they would. It is going well with the federal government – the talks are going on, not about feed-in tariffs, but about other necessary measures; but with your help it is going even better.”

Many jobs would be created; above all direct jobs on site, through CSP more than through PV or wind energy, and many indirect jobs through spin-offs. Through project funding, costs could soon fall to market levels. "Then we have done our job, for the next 10 to 15 years the politicians must stimulate this." To the terrorist question he replies: "If I were a terrorist, I'd attack fossil hydrocarbon first. (Laughter). In addition, no politically sensitive areas such as Western Sahara will be visited." That is exactly what Koczy had asked...

A good hour later Paul van Son is back on the plane to Munich.

3. A brief outline of the history of renewable energies

The sun has preoccupied people since the very beginning of their existence. For life on the blue planet, it is the most important celestial body and was of considerable importance even before history was written. In ancient times she was elevated to the status of a god by people around the globe (Petermann 2008). For the Sumerians she was a sun god, according to the spiritual ideas of the Babylonians she entered the sky every day, also as a sun god – Shamash. In their traditions they suggest that nothing remained hidden from the rays of their sun god. The Egyptians worshipped the sun god Ra. In ancient China, the sun stood for Easter, spring, birth, masculinity and the emperor. Greece at that time worshipped the sun god Helios, who found his counterpart in the Roman god Sol Invictus. The ancient advanced civilizations in Central and South America also paid homage to sun gods. And the orientation of the cult site in Stonehenge can probably be traced back to the movement of the sun. Scientists such as Copernicus, Galilei and Newton shaped our present image from the end of the Middle Ages, in which the sun is at the centre of the universe and the planets orbit it. This led to considerable conflicts with the church.

First findings about the use of the energy emanating from the sun also date back far B.C. According to tradition, in Mexico the Omeks, who were later displaced by the Mayan culture, used parabolic mirrors made of magnetic iron as lighters.

The first developments in southern Europe can be traced back to Archimedes. The mathematician is said to have developed burning mirrors with which it was possible to ignite distant ships. The sun's rays had to generate temperatures of 300 degrees Celsius over long distances in order to ignite wood. Not an easy task, as researchers at the Massachusetts Institute of Technology (MIT) discovered at the beginning of the century.

In the 18th century, the natural scientist Horace-Bénédict de Saussure developed the precursor of today's solar collectors. This consisted of a simple wooden box with a black bottom and a glass cover. Even so, the water in these simple collectors reached a temperature of 87 degrees Celsius.

About 100 years later, in 1866, the French mathematics teacher Augustin Mouchat developed the first solar steam engine (Mouchat 1877). Using a concave mirror, it concentrated water in a glass cylinder. Solar energy heated the water, so that machines could be driven by the expanding steam. The solar thermal pioneer continued to develop his invention, for which he was awarded a gold medal at the Paris World Exhibition in 1878. One year earlier, the French government had already expressed its interest in the system for deployment in its colonies but had rejected the plan.

These last decades of the 19th century were a time of upheaval, driven by the development of the steam engine and industrialisation. The new machines in Europe and America were fed by coal. The course was set for a fossil energy supply. As a social consequence, a working class on an unprecedented scale emerged in Europe, causing corresponding tensions. Karl Marx wrote *Das Kapital* in England and the Social Democratic Party was formed in Germany. A reaction to this was the social legislation of German Chancellor Bismarck.

In addition to the path taken in the direction of a fossil raw material supply, the focus was also on solar energy from time to time. One of the founders of social democracy, August Bebel, quoted the Leipzig physicist Kohlrausch in his book *Frauen im Sozialismus* (Women in Socialism), published in 1870 and supplemented by the chapter on Electroculture in 1900: "The parts of the earth's surface, and especially the largely unused parts, to which the sun's heat flows so regularly that it can also be used for regular technical operation, offer a wealth of energy that far exceeds all requirements. Perhaps it would be a good idea for a nation to secure a share of such areas now. Very large areas are not even necessary; a few square miles in North Africa would suffice for the needs of a country like the German Reich. A high temperature can be generated by concentrating the heat of the sun and with this transportable mechanical work, accumulator charges, light and heat can be produced, or by electrolysis also directly burning material." (Bebel 1879).

At the turn of the century, physics was on the verge of a tremendous leap in development. Introduced by the Age of Enlightenment in Europe, outstanding processes of knowledge in the natural sciences took place towards the end of the century. Wilhelm von Humboldt's education policy bore fruit, and in Germany in particular the higher education landscape had developed well. Until 1918, researchers from German universities received more than a third of all Nobel Prizes for scientific work. At that time, neighbouring Great Britain had a dynamic innovation centre where students from all over the world were educated. Electromagnetic waves were investigated in Munich, Bonn, Halle and Göttingen, Berlin, Breslau as well as Würzburg. The speed of light could already be determined, the electric current still remained mysterious.

On the industrial side, Werner von Siemens developed a considerably improved dynamo in 1867, driven by steam engines or water turbines (Dettmann et al. 2013). The development of electric motors began around 1890. They succeeded in taming the current by transporting it from Lauffen am Neckar to Frankfurt am Main – over 175 kilometres after all. At the 1891 World Exposition, the power line spectacularly demonstrated the efficiency of electricity and this is regarded as the birth of electrical energy supply. After the World Exhibition, the demand for electrical energy increased rapidly. The incandescent lamp was as successful against gas light as the electric motor against the steam engine with transmission.

In Berlin, the physical-technical Reichsanstalt developed modern precision technology. At that time, the German capital was one of the world centres for physics (Schroeder 2004). Max Planck achieved a decisive breakthrough. He discovered that energy does not flow continuously but is transferred in packets. For the calculation he defines the natural constant h. The world view of physics was turned upside down. During this time, the French physicist Alexandre-Edmond Becquerel observed the interactions between light and electrons. Albert Einstein also described this effect theoretically by considering the quantum theory of light. These were decisive findings for the later understanding of photovoltaic modules.

The still young German Reich wanted to achieve world fame through scientific and technical pioneering achievements. The art of engineering was highly valued by the public. The Ruhr area was developed in Germany as the industrial centre of its time. This is where Rheinisch-Westfälische Elektrizitäts AG, RWE, was founded at the turn of the century. Their recipe for success was based on the rapid conversion of the hard coal extracted on site into electricity. This stormy development of coal as a primary energy source around the turn of the century has been accompanied by repeated references to renewable alternatives: "Not too distant is the day when the exploitation of the sun's rays will revolutionise our lives, man will free himself from dependence on coal and hydroelectric power, and all large cities will be surrounded by enormous apparatuses, real solar radiation traps, in which the solar heat is collected and the energy gained is accumulated in mighty reservoirs." The American inventor Thomas

Alva Edison pointed out solar energy: "I would invest my money in solar energy – what an energy source! I hope we don't have to wait for oil and coal to run out until we figure this out."[2]

The American inventor Frank Shuman had already experimented with a solar-heated, ether-powered toy steam engine in 1897 (Dettmann et al. 2012). Around the same time, in 1907, Wilhelm Meier from Aalen and Adolf Reneshardt from Stuttgart applied for a patent for a device for the direct utilisation of solar heat for the purpose of steam generation in parabolic trough collectors. It took another five years until, from 1912 to July 1913, under the leadership of Frank Shuman, the first parabolic trough power plant was built in Maadi, Egypt, about 25 kilometres south of Cairo. Charles Boys, an English physics professor who had criticised his design, engaged Shuman as his advisor. They used parabolic trough collectors with a length of 62 meters, an aperture width of 4 meters and a total aperture area of 1,200 square meters. "20,000 square miles of collectors in the Sahara," wrote Shuman, "could permanently provide the world with 270 million horsepower" (*New York Times* 02.11.1916). Shortly before the First World War, the German Reichstag approved 200,000 Reichsmark for a parabolic trough demonstration in German Southwest Africa. The First World War intervened, then the oil age and the discovery of fossil raw materials in the Middle East prevented further implementation of the plan.

Solar thermal power plants only became interesting again at the end of the 1970s. Due to the oil crisis, energy prices had shot up and alternatives were sought. Electricity supply companies undertook to purchase electricity from independent producers at clearly defined costs. The Californian utility Southern California Edison (SEC) agreed to this by offering long-term feed-in conditions. In 1983, the newly founded company LUZ International Ltd.

---

2 http://www.entri-consulting.com/de/energie.

signed a 30-year contract with the SEC to feed solar power into the grid. In 1984, the first solar thermal parabolic trough power plant was built in the Mojave Desert in California. By 1991, power plants with a total capacity of 354 megawatts had been installed on a land area of over seven square kilometres. The power plants could also be operated with fossil fuels during the night hours and in unsuitable weather conditions.

At the beginning of the century, the market for thermal solar power plants continued to develop in Spain and the USA. German companies in particular were responsible for project planning and the supply of components. Schott Solar AG produced absorber tubes for the focal point of the parabolic trough mirrors, Siemens AG from Munich produced the turbines, and Flabeg from the Bavarian forest supplied the mirrors. At the beginning of the century, Spain provided a feed-in tariff for thermal solar power plants: Around 27 eurocents were paid in the southern European country for each kilowatt-hour of solar heat electricity fed into the grid. This tariff has been guaranteed for 25 years. Erlangen-based Solar Milennium AG built a total of three parabolic trough power plants, Andasol I, II and III, in Andalusia in southern Spain. The financing volume of the Andasol power plant amounted to 1,300 million euros. These solar power plants were each equipped with two gigantic storage tanks in which 57,000 tonnes of liquid nitrate salt were melted to 400 degrees Celsius for energy storage. The solar thermal power plants achieved efficiencies of up to 15 percent. In addition, a considerable amount of maintenance was required: the mirrors have to be cleaned regularly. The electricity generation costs achieved by solar thermal power plants were still too high to displace conventional power plants under the conditions of a free market (Scherer/Stolten 2013).

Compared to solar thermal power plants, the time was not ripe for the development of photovoltaic modules until later

(Eschrich/Wagmann 2007). Here they needed the new realizations of physics. They go back to the Lebanese inventor Hassan Kamel Al-Sabbah. Al-Sabbah is responsible for over 70 patents, most of which deal with tube technology in televisions. However, his work also includes the production of the first solar cells. The Lebanese first studied electrical engineering and mathematics at the American University of Beirut and in Damascus, Syria. In 1921 he emigrated to the USA, where he continued his studies at the Massachusetts Institute of Technology (MIT) and the University of Illinois. This makes him an early harbinger of his Arab compatriots who are aiming for training at foreign universities.

Wind power, too, can look back on a long history of development (Gasch/Telwe 2013). According to historians, the Orient is the cradle of wind energy. First indications can be found already in the year 1700 B.C. in Mesopotamia, present-day Iraq and Syria. Windmills are said to have irrigated the plains there. There is also documentary evidence of an early use of wind power in present-day Afghanistan. In China, the development of wind turbines began around 1000 A.D. In wind turbines of the horizontal type that were initially developed, braided mats for air resistance were attached around the vertical axis of rotation. The air flow caused the system to rotate. With this vertical rotation principle, the grinding stone could be driven directly. In the Occident, the wind current was transferred horizontally to the rotation axis. The classic of the past centuries is the post-mill. It was widespread in England, France, Holland, Germany and Poland. It consists of a box-shaped mill house which is mounted on a trestle so that it can rotate around a pin. This allows it to be turned into the wind together with an impeller. It was used exclusively for grinding. From the 12th to the beginning of the 19th century, hydroelectric and wind power were used to generate mechanical energy. A centre for this technology was Holland; in the second half of the 19th century there were about 9,000 windmills. They were used for grinding grain, sawing, hammering and pumping water.

Figure 41 A typical windmill in the Netherlands. Source: Thomas Isenburg

This was also the time when James Blyth from Scotland first converted wind into electrical energy. The scientist was an electrical engineer at various Scottish universities and a pioneer in the field of power generation from wind energy. The wind turbine he built was used to light his holiday home in Maryhill, the improved version was used to supply a hospital. However, the advancing industrialization displaced windmills by fossil-fuelled aggregates; after the First World War, only science was concerned with them for the time being. With the reconstruction of the destroyed Europe after the Second World War and the awareness that coal reserves would soon dwindle, a renaissance of technology began. The oil price shock from 1973 to 1978 gave new food to the social debate about an alternative energy supply. At the beginning of the 1980s, the Danish agricultural machinery manufacturer Vestas successfully began building smaller wind turbines. They had a rotor diameter of 12 to 15 meters. Previously, huge wind turbines had been developed by the aerospace industry with government subsidies in the USA, Germany, Sweden and some other countries, almost all of which failed after a few hours of operation. Renewable energy sources supply the electricity in a fluctuating manner. This is why experts worldwide were discussing the conversion of renewable electricity into hydrogen as an energy carrier. This could then cover the continuous demand. The first references to the use of reactive gas as an energy carrier can be found in the novel by the French science fiction author Jules Verne from 1874, *The Mysterious Island*. Here Verne writes: "Yes, my friends, I believe that one day water will be used as fuel, that hydrogen will be used as an inexhaustible source of warmth and light." In the period that followed, concepts for the energy-economical use of gas repeatedly appeared alongside its extensive use as a raw material for chemistry. However, hydrogen in a mixture with oxygen is extremely reactive. The energy stored in the molecules is released by the famous oxyhydrogen reaction. The development of storage technology suitable for everyday use has lasted until the recent past. In the meantime, the problem has been solved reliably.

The advantage of hydrogen as a storage medium lies in its cost-effective, flexible application. In terms of process technology, it is easily accessible via a classic electrolysis of water with the help of renewable electricity. Fuel cells can be used to generate electricity from hydrogen and oxygen. Increasing the efficiency of processes is the subject of extensive research worldwide. In addition to Europe and the USA, Southeast Asia and Japan are at the centre of activities.

Hydrogen chemistry has a long tradition in Germany. A milestone is the development of the Haber-Bosch process. It was developed by the German chemists Fritz Haber and Carl Bosch at the beginning of the 20th century and is the central step in ammonia synthesis from atmospheric nitrogen and hydrogen at pressures of about 150 to 350 bar and temperatures of about 400 to 500 degrees Celsius. It is also used for the synthesis of ammonia from nitrogen. Ammonia is the basis for the production of fertilizers and explosives.

After aircraft pioneer Bölkow had supplied ideas for hydrogen production in deserts with renewable electricity, the German Aerospace Center (DLR) was concerned with hydrogen production in the desert. For this purpose, a 350-kW electrolyser operated with solar cells produced proof that the production of storable and transportable hydrogen was possible. The available solar resources could supply one percent of Saudi Arabia's land area with the same amount of energy as is exported annually as crude oil.

These thoughts fit into the ideas of the time with the development of the Desertec project. However, this idea is not unrivalled. In his book *Beyond Oil and Gas: The Methanol Economy*, the American chemistry Nobel Prize winner with Hungarian roots and scientific giant Georg A. Olah proposed a hypothetical methanol economy in 2005. Methanol is the simplest alcohol and can already be added to fuel today. The liquid can be transported much more easily than a gas. Ideally, carbon for methanol synthesis could in future be obtained from atmospheric rather than fossil sources with the aid of renewable energies. However, the chemist, who died in 2017, also included nuclear power as an energy source.

The chemical recycling of carbon dioxide from nuclear power plants could be integrated into this process.

The Dutch process specialist from TU Delft, Ad van Wijk, provides arguments for a hydrogen economy. His considerations begin with a global hydrogen cycle. For the individual steps, the Dutch scientist then makes detailed considerations with the aim of exploring the possibilities of hydrogen as an energy carrier for the future. The gas will be obtained at the locations through renewable electricity which has high potential for regenerative power plants. This can be the MENA region with its possibilities for solar and wind power, but also offshore wind farms in the North Sea. His vision is to transport the gas in ships to energy ports. Here the conversion into electricity takes place via a fuel cell or feeding into the gas grid. Gas caverns offer another storage option for storing large amounts of energy.

4. The Europeans in North Africa – An insight into recent history using Morocco and Algeria as examples

Solar thermal power plants are designed for locations with high direct solar radiation. Places in the equatorial sun belt of the earth up to geographical latitudes of about 30 to 40 degrees north or south are particularly suitable. In Europe there are favourable conditions in Spain, the MENA countries occupy this corridor within reach of Europe. In recent decades, however, these countries have had to endure many conflicts, mostly involving the replacement of colonial powers, raw materials and religion.

Towards the end of the eighteenth century, the old kingdoms of the Muslim world were in decline, while Europe was strengthening. Then, in the nineteenth and twentieth centuries, a new world order was gradually imposed on the Arabs until they could finally free themselves (Hourani 2016).

Shortly before, in 1764, James Watt had considerably improved the existing steam engines and the industrial revolution began. A little later, in 1776, the United States gained its independence. The French Revolution broke out in Europe in 1789. The situation began to change dramatically. There was a gap between the technical performance of the European countries and the rest of the world. In the period that followed, factories spread and steam-powered means of transport such as railways and steamboats crossed the landscapes and seas. Messages could be transmitted via telegraph lines. European trade really took off as a result of these technical achievements.

The Arab world was largely occupied by the Ottoman Empire. In the east, this reached as far as Iran. In the north and west there were Christian states. The Ottomans still controlled the east and south coasts of the Mediterranean. Morocco at the western end was dominated by the Alawites. They attribute their lineage to the Prophet Mohammed and Morocco's current king, Mohammed VI, represents him today.

There had been no technical progress in this region on the southern and eastern coasts of the Mediterranean and the level of scientific knowledge had even fallen. It was in decline compared to Europe and lost parts of its cultural heritage.

Meanwhile, trade between the Christian and Muslim worlds increased: sugar from the Caribbean, French textiles, silk from Lebanon, cotton from Northern Palestine, cereals from Algeria and animal skins from Morocco were exchanged. The players in this market came from Venice, Genoa and later also from England and France. This had only a limited impact at first but was a sign of a shift in power.

The expansion of Europe then became clear between 1830 and 1870, with steamers from London and Liverpool, Marseille and Trieste regularly calling at the ports of the eastern and western Mediterranean. The Ottoman government tried to learn from the Europeans. It copied methods for building up the military, administration and legal system. In the port cities and capitals of the Ottoman Empire there was a lively trade with Europe. In addition to the ruling elite, politics favoured those merchants who traded with Europe. The exchange of goods through import and export grew and merchants gained an ever-greater influence on the market. They also determined production by making capital available to landowners or farmers and deciding what should be grown. The most successful traders were Europeans because they were familiar with the European market and had access to bank loans at all times. Ottoman Christians and Jews, Greeks and Armenians, Syrian Christians from Baghdad, Tunis and Fez also participated in the great business deals with Europe. They knew the local markets and were well suited as middlemen.

Islam remained the religious and legal basis. However, Muslims tried to find reasons for Europe's strength. They wanted to adopt their successful strategies and the ideas behind them. Previously, representatives of this current of Islamic society had visited mission schools. They were able to disseminate their ideas to the public thanks to new media such as magazines and newspapers.

After some time, first Egypt, then Tunisia, Morocco and Libya came under European control. This was mostly preceded by an immense indebtedness of the Arab countries to European banks. At the end of the First World War the Ottoman Empire finally broke up and Turkey emerged from the ashes.

Until decolonisation, Europeans had divided Africa among themselves. In North Africa, Morocco, Tunisia and Algeria belonged to the French-dominated part of this world order. Libya was occupied by the Italians and Egypt was dominated by the British. A small part of present-day Morocco was also occupied by redevelopers. The Berbers in central Morocco, who mainly inhabit the High and Middle Atlas, are divided into 300 tribal groups. In the cities of the north , Arab Berbers dominated the population. Over the centuries, they have mixed with the immigrant Arab population and largely adopted their culture and language. Few inhabitants of Morocco are pure Arabs, the small group called Hartians are descendants of slaves or mercenaries from the 11th century.

In the Treaty of Fez, which regulated the subordination of central Morocco to France, Morocco was divided into several zones. Spain received two strips of land – one in the south , the other on the Mediterranean coast. Morocco was thus part of the French protectorate from 1912 to 1956. Sultan Mulai Yusuf (1912-1927) and King Mohammed V (1927-1957) called themselves Moroccan heads of state, but their power was limited. It was actually exercised by the Europeans: Military violence was in their hands and they dominated the political processes. The guidelines for personnel, economic and foreign policy are also dictated by Spaniards and French. Again and again, the various Moroccan tribes resisted the European occupation. The tragic climax was the initiative of King Alfonso VIII of Spain, who planned the extermination of the Rif Kabyles, a Berber tribe living on the Rif Atlas and the Tangier coast, as a retaliatory measure. He bought poison gas from the Stoltzenberg chemical factory, which was jointly responsible for the poison gas attacks in the First World War. The chemical weapon 'Lost' or mustard gas was used for the first time from the air and also specifically against the civilian population. France limited itself to the use of tear gas and did not oppose the Spanish approach. Great Britain also sacrificed the question of justification to its Mediterranean interests, as Dirk Sasse elaborated in his dissertation *Franzosen, Briten und Deutsche im Rifkfrieg 1921-1926. Spekulanten und Sympathisanten, Deserteure und Hasardeure im Dienste Abdelkrims* (Frenchmen, Brits and Germans in the Rif War 1921-1926. Speculators and sympathizers, deserters and gamblers in the service of Abdelkrim) (Sasse 2006). As early as 2002, the *Tageszeitung* (TAZ) wrote about the devastating long-term consequences: "In the cancer centre in the Moroccan capital Rabat, 60 percent of patients come from the former war zone (...). Many are descendants of war victims." (Wandler 26.01.2002). During the use of chemical weapons, all three European countries worked on the protocol of the League of Nations, which prohibits the use of poison gas.

In the 1930s, the anti-colonial resistance began to look for new forms of organization. Until 1943/1944, a central organization was formed from which the Party of Independence was eventually founded. In 1951/1952, resistance to the French military's repression escalated to open war and a general strike paralysed the country. In 1953 Mohammed V was deposed as ruler and his family was sent into exile to Madagascar. This in turn triggered an escalation. In the meantime, negotiations on the country's independence were conducted with the Europeans.

After the death of his father, Hassan II took the throne in March 1961 and pursued a conservative policy. He relied above all on the old elite and feudal structures of the country. He was not popular and was assassinated before his accession to the throne. First Morocco was a constitutional monarchy, then the monarch dissolved parliament in 1965 and took over the affairs of government alone. Especially left-wing opposition members were persecuted. The Algerian-Moroccan border war took place in 1963. Hassan II survived two republican coups, in 1971 in the Moroccan Skhirat and in 1972, only barely, on his return from Paris. Human rights were then trampled underfoot, all kinds of opposition persecuted. People disappeared and were tortured.

In 1975, Hassan II initiated the Green March. Some 350,000 unarmed Moroccan civilians occupy the former Spanish colony of Western Sahara in the south of the country. Soon thereafter the military resistance of the locals began under the Polisario. This led to tensions with Algeria.

One of the positive impulses provided by Hassan II was the systematic construction of numerous dams to supply the local agricultural industry with water. He also advocated a solution to the Israeli-Palestinian conflict. At the beginning of the 1990s, under great domestic and foreign political pressure, he began a slow liberalization of politics. In 1999 Hassan the II died and his son Mohammed VI became king. At the end of the reign of Hassan II, Morocco was heavily indebted, the illiteracy rate in the country was 50 percent and many of his compatriots were bitterly poor.

Another development took Algeria. The middle one of the Maghreb countries is the largest territorial state in Africa. France had a dominant influence on developments, from the conquest in 1830 to 1847, when the Muslim Algerians called for jihad. Until the final conquest, the Algerians suffered massive losses. France described the occupation of Algeria as an attack by Christianity on Islam. It was the first major conquest of an Arab country and France's oldest and largest colony.

In 1848 the North African country became a territorial part of the Grande Nation. In fact, Algerians and French settlers, known in France as Pieds-Noirs, did not have the same rights. For the Arabs, this defeat meant that they could no longer live in a stable and independent cultural unity. Algeria's religious structures were shattered and the common property of the Muslim communities on the land was abandoned. Around 1865 numerous settlers came to the country: Italians, Spaniards, Maltese and French. It was the policy of the French to strengthen their influence in the country through settlement programmes in the sense of a *Colonie de peuplement*. The population was divided into two classes: Frenchmen and subjects. Until the Second World War, the French acquired more and more farmland. As early as 1936, they owned 40 percent. At the same time, the number of Algerians grew from 2 million to 9 million by the outbreak of the Second World War. In the 100 years of French rule, the Algerians had considerably impoverished. Malnutrition and famine were part of their everyday lives. Muslims were also excluded from education. During the Second World War, the Algerians who fought alongside France were promised independence. After the war, the Democratic Union of the Algerian Manifesto (UDMA) party was founded. It followed the policies of Atatürk, the founder of modern Turkey, and rejected Arabism. The Algerians called on the Europeans to give up the pride of conquest and their colonial complex. Even after the end of the Second World War, the North African country, which was affiliated to France, had the right to vote in two classes, which distinguished between Algerians and French. This caused unrest and the resistance formed.

The National Liberation Front of Algeria (FLN) was founded in 1954. It advocated open resistance to French rule. At first it stood for a socialist democratic society according to Islamic principles. The armed struggle was the means to achieve independence from France. It saw itself as the only legitimate political representation of the non-European population and asserted this claim by force against more moderate political forces. Their headquarters was the Egyptian capital Cairo. Preferred targets of the insurgents among the locals were Qaids, elected representatives at the municipal level, tax collectors, taxpayers, and Algerian members of the French armed forces. It came to a guerrilla war with horrible consequences. Victims were also the pieds-noirs and more moderate Muslims. In May 1958, French Algeria was then founded after the general government had been established in the course of a demonstration. DeGaulle took over the presidency in the same year. He prevented Algeria from voting on its own form of government and its relations with France. The many French settlers in the country felt ignored and founded the French National Front. During a large-scale demonstration organized by them 26 people were killed. By declaring the Algerian war an internal French conflict, the government in Paris bypassed the 1949 Geneva Convention on the Protection of Civilian Persons, which elicited strong international criticism. In 1962 the FNL was effectively defeated. There were peace negotiations that led to a referendum. The land was devastated and there were hundreds of thousands of victims. In the 1961 referendum, almost 60 percent of Algerians and 75 percent of French voted for independence from mainland France.

The FLN candidate Ahmed Ben Bella was elected president of Algeria in the first elections in Algeria. His election promise was a revolutionary Arab state based on the principles of socialism. To this end he installed a collective leadership at home and an anti-imperialism abroad. Ben Bella was overthrown by Houari Boumedienne in a coup d'état on 19 June 1965. Under Boumedienne's leadership, the country's mineral resources, in particular oil and gas, were increasingly exploited to promote the country's industrial development and Algerian socialism. He nationalised the oil industry and parts of agriculture and tried to protect the burgeoning industry with high tariffs. The 1976 constitution confirmed Boumedienne's position of power and the elimination of all opposition.

After his death in 1978, he was replaced by Colonel Chadli Benjedid, who was re-elected in 1984 and 1989. He was the Algerian president until 1992. During this time there was hardly any political change and or signs of opposition. Food in the resource-rich country became scarce in 1988 and thousands protested. Between 160 and 600 people were killed. The government gave in and wanted to pave the way for a multi-party system.

This resulted in a strengthening of the Islamists. Already in local government elections in spring 1990, the previously banned fundamentalist party (FIS) won a landslide victory. In the subsequent parliamentary elections in 1991, the FLN, which had previously been the ruling party, won only 15 of the 231 seats. The clear winner was the FIS. At this point the army intervened again. The second ballot was cancelled and the FIS leaders were arrested or fled into exile.

There was another war in Algeria. This time it was a civil war. New elections in 1995 did not improve the situation, Islamic parties were not admitted. The government emerged victorious in the election. However, this election victory was accompanied by strong accusations of bribery, so that the war was waged even more brutally. The Groupe Islamique Armé (GIA), which committed acts of violence and atrocities during the Algerian civil war, felt provoked by the French aid for the Algerian government. That is why there were bombings and kidnappings in France. In the conflict with the Islamists, state security forces finally gained the upper hand. New elections in April 1999 saw former Foreign Minister Abdelaziz Bouteflika as the winner.

However, the opposition had withdrawn from the election under accusations of fraud.

For the French, Algeria is a difficult chapter. The Algerian war was not recognised as such until 1999, when France was a great nation with a great history. Algeria advocates enlightenment and the introduction of human rights and can thus establish a proud identity. Perhaps this is the very reason why processing takes longer. Since the end of the Algerian war, no French president has dealt with the trauma of colonisation. However, the left-wing liberal Emmanuel Macron has a new approach. On a visit to Algeria, he described colonisation as a crime against humanity, not without provoking great indignation in France. Jean Marie LePen of the Front National described the Algerians living in France as the riffraff of the suburbs. There is suffering on the other side as well. About 3.2 million Pieds-Noirs and their descendants live in France. They had to flee the country upon Algeria's independence. Up to the present day, Algeria's recent history has had considerable consequences.

5. Legal systems and developments in Saudi Arabia and the United Arab Emirates

The Arab world is characterized by legal structures that are little known outside the region and often can only be understood with great effort. This book does not intend to give an in-depth description of the 'legal structures', but it does want to give some insight into two examples.

There are various parallels between the legal systems of the countries in the Middle East and those of European or Western countries. For instance, states have constitutions, which contain the fundamental rules of the community and grant citizens fundamental rights, and courts are responsible to guarantee the enforcement of these rights and to ensure the enforcement of rights between citizens. A constant change and evolvement of legal systems can be observed due to – in particular – an increased exchange and global cooperation between various companies from different countries. For the purposes of this article, the legal systems of the Kingdom of Saudi Arabia and the United Arab Emirates have been chosen to exemplify the legal systems of the Middle East region.

Foray

The legal system and developments in the Kingdom of Saudi Arabia

Author: **Murad M. Daghles** Chancellery: **White & Case LLP**

1. *General information*

The Kingdom of Saudi Arabia is an absolute monarchy and the legal system of the Kingdom of Saudi Arabia is based on the *Shari'ah*.[3] The two main sources of *Shari'ah* are the *Qur'an* and the writings detailing the Prophet Mohammed's saying and actions (the so-called *Sunnah*). Based on the "Basic Law of Governance"[4] of 1992, Article 1 states that the *Qur'an* and the *Sunnah* constitute the constitution of the Kingdom of Saudi Arabia.[5] *Shari'ah* has four main schools of jurisprudence: Hanbali, Hanafi, Shafi and Malaki.[6] Generally, courts in the Kingdom of Saudi Arabia apply the Hanbali school of jurisprudence.[7]

2. *The judicial system*

The judicial system in the Kingdom of Saudi Arabia is undergoing constant reform and modernization.[8] The judicial system

3 https://www.saudiembassy.net/legal-and-judicial-structure-0,

4 The Basic Law of Governance, issued pursuant to a Royal Order No. A/90, (27/8/1412H, March 1st, 1992).

5 Ansary, *A Brief Overview of the Saudi Arabian Legal System*, 2015, sec. III., http://www.nyulawglobal.org/globalex/Saudi_Arabia1.html,

6 Daghles, *Die Kompatibilität islamischer Staatsauffassungen mit der freiheitlich demokratischen Grundordnung*, p. 112.

7 Ansary, *A Brief Overview of the Saudi Arabian Legal System*, 2015, sec. IV.3.8, http://www.nyulawglobal.org/globalex/Saudi_Arabia1.html,

8 Ansary, *A Brief Overview of the Saudi Arabian Legal System*, 2015, sec. V., http://www.nyulawglobal.org/globalex/Saudi_Arabia1.html,

consists of three parts: The first level courts are the main part. They consist of several instances. First instances include criminal courts, family courts, labour courts and commercial courts.[9] Following the first instance there are 'Courts of Appeal'[10] and 'Supreme Courts'[11]. The second level courts are the so-called 'Board of Grievances', which deal with cases involving governmental actions.[12] The third level are quasi-courts, which consist of committees within governmental entities and deal with specific disputes, such as securities disputes.[13]

3. *Legislature*

a) Royal Decrees

Royal Decrees primarily serve as a legislative instrument for enacting (among other things) laws, treaties, international agreements, and concessions.[14] However, Royal Decrees can also be issued in connection with other non-legislative matters (such as forming entities and appointing certain government officials).

Three basic laws encompass the authority and process for the issuance of Royal Decrees. The Basic Law of Governance[15] sets out the general framework under which Royal Decrees are issued. The Law of the Council of Ministers[16] and the Law of the

9 Art. 9 para. 3 Law of Judiciary, issued pursuant to a Royal Decree No. M/64, (14/7/1395H, July 23th 1975).
10 Art. 9 para. 2 Law of Judiciary.
11 Art. 9 para. 1 Law of Judiciary.
12 Ansary, *A Brief Overview of the Saudi Arabian Legal System*, 2015, sec. IV.3.2.2, http://www.nyulawglobal.org/globalex/Saudi_Arabia1.html,
13 Ansary, *A Brief Overview of the Saudi Arabian Legal System*, 2015, sec. IV.3.2.3, http://www.nyulawglobal.org/globalex/Saudi_Arabia1.html,
14 Art. 70 Basic Law of Governance.
15 Law of Grievances, issued pursuant to Royal Decree No. M/78 (19/9/1428, October 1st, 2007).
16 Law of Council of Minister, issued pursuant to Royal Order No. A/13 (3/3/1414H, August 21st, 1993).

Shura Council[17] further set out the roles of each of the Council of Ministers and the *Shura* Council, including in connection with the process of issuing Royal Decrees.

b) Royal Orders and Resolutions of the Council of Ministers

In addition, legislation in the Kingdom of Saudi Arabia is also enacted through various other legislative instruments such as Royal Orders and Resolutions of the Council of Ministers. A Royal Order is typically regarded as the highest instrument of governance in the Kingdom of Saudi Arabia, which, despite serving primarily as an executive instrument, has ultimate legislative and executive power. Royal Orders are issued directly by the King and do not require the approval of the Council of Ministers, the Shura Council, or any other legal body, unlike Royal Decrees which typically require the pre-approval of the Council of Ministers and, in certain cases, the Shura Council.

Resolutions of the Council of Ministers are issued in connection with various legislative and non-legislative matters. The Council of Ministers issues its resolutions either as part of the process of issuing a subsequent Royal Decree or as independent decisions.

4. *Legal bodies*

a) The Council of Ministers

The Council of Ministers is a legal body whose meetings are chaired by the Prime Minister or his deputies.[18] The Council of Ministers is comprised of (1) the Prime Minister, (2) Deputies of the Prime Minister, (3) Ministers, (4) State Ministers, and (5) advisors to the King (appointed from time to time by way of a Royal Order).[19]

17 Law of *Shura* Council, issued pursuant to Royal Order No. A/91 (27/8/1412H March 2nd, 1992).

18 Art. 7 Law of the Council of Ministers.

19 Art. 12 Law of the Council of Ministers.

A meeting of the Council of Ministers is only considered valid with a quorum of at least two-thirds of its members, and resolutions are passed by a majority vote. In the event of a tie, the Prime Minister has a casting vote. In exceptional cases (determined by the Prime Minister), meetings of the Council of Ministers are considered valid if only half of the members are present. In such cases, however, resolutions are not considered valid without the approval of at least two-thirds of the members present.

The Council of Ministers has a broad role of formulating the internal, external, financial, economic, educational, and defence policy as well as all the general affairs of the Kingdom of Saudi Arabia and supervising the execution of the same.[20] It issues resolutions in connection with matters involving the enactment of laws, regulations, treaties, international agreements, and concessions, which are issued as part of the process of issuing a subsequent Royal Decree. It also issues resolutions in connection with other legislative and non-legislative matters either as part of a Royal Decree process (such as forming entities) or as independent decisions (such as establishing certain government agencies and issuing supplementary regulations). Resolutions of the Shura Council are reviewed by the Council of Ministers.[21]

Each Minister has the right to propose bills of law or regulations relating to the affairs of his ministry and any member of the Council of Ministers may propose items for discussion at the meetings of the Council of Ministers, subject to the Prime Minister's approval.[22]

b) The Shura Council

The Shura Council is a consultative legal body comprised of a chairman and 150 members, all of whom are appointed by way of

20 Art. 19 Law of the Council of Ministers.
21 Art. 19 Law of the Council of Ministers.
22 Art. 22 Law of the Council of Ministers.

a Royal Order.[23] A meeting of the Shura Council is only considered valid with a quorum of at least two-thirds of its members, including the chairman or his deputy. Resolutions are passed by a majority vote and, in the event of a tie, the chairman has a casting vote.

The Shura Council provides its opinions regarding public policy if so requested by the Prime Minister. The Shura Council has a broad consultative mandate relating to various affairs, including but not limited to:[24]

- reviewing and commenting on the general economic and social development plan, laws, regulations, international treaties and agreements, concessions, and annual reports submitted by ministries and other governmental bodies;
- interpreting laws; and
- proposing bills of law or amendments to enacted laws.

The Shura Council serves primarily as a legislative body, and accordingly matters involving the enactment of laws, certain regulations, international treaties and agreements, and concessions are subject to the Shura Council's review. Non-legislative matters (such as forming entities) are typically not referred to the Shura Council, unless otherwise requested by the Prime Minister.

5. *Developments*

Since 2000, there has been a progressive liberalization of the legal system of the Kingdom of Saudi Arabia, which has not yet come to an end.[25] There has been a steady increase of the population's participation rights. A landmark was the Royal Decree of

23 Art. 3 Law of the *Shura* Council.
24 Art. 15, 18, and 23 Law of the *Shura* Council.
25 Keimer, in *Länderreport Saudi-Arabien*, 126, 126.

King Abdullah of 19 October 2006.[26] It established the Pledge of Allegiance Commission, which is responsible for the election of the Crown Prince.[27] In 2012, women were granted both the right to vote and to stand for election.[28] In March 2015, King Salman revealed further comprehensive reform plans, following his predecessors' footsteps.[29] He also called for greater participation of the population in decision making and for the fight against corruption and social injustice.[30]

B. The legal system and developments in the United Arab Emirates

1. General information

The United Arab Emirates (UAE) is a presidential, federal state comprised of seven emirates: Abu Dhabi, Ajman, Dubai, Fujairah, Ras Al Khaimah, Sharjah and Umm Al Quwain.[31] Each of the seven emirates has a high degree of autonomy and its own public service.[32] Abu Dhabi and Dubai, for instance, have large authorities for economic and public affairs.[33] The areas of

26 Ansary, *A Brief Overview of the Saudi Arabian Legal System*, 2015, sec. II.2., http://www.nyulawglobal.org/globalex/Saudi_Arabia1.html,
27 Ansary, *A Brief Overview of the Saudi Arabian Legal System*, 2015, sec. II.2., http://www.nyulawglobal.org/globalex/Saudi_Arabia1.html,
28 Ansary, *A Brief Overview of the Saudi Arabian Legal System*, 2015, sec. II.2., http://www.nyulawglobal.org/globalex/Saudi_Arabia1.html,
29 Ansary, *A Brief Overview of the Saudi Arabian Legal System*, 2015, sec. II.3., http://www.nyulawglobal.org/globalex/Saudi_Arabia1.html,
30 Ansary, *A Brief Overview of the Saudi Arabian Legal System*, 2015, sec. II.3., http://www.nyulawglobal.org/globalex/Saudi_Arabia1.html,
31 https://www.auswaertiges-amt.de/de/aussenpolitik/laender/vereinigtearabischeemiratenode/vereinigtearabischeemirate/202316, last visited on March 21, 2019.
32 Seifert, *Rechtliche Rahmenbedingungen für Geschäftstätigkeiten in den Vereinigten Arabischen Emiraten*, 4th edition, 2014, p. 41.
33 Seifert, *Rechtliche Rahmenbedingungen für Geschäftstätigkeiten in den Vereinigten Arabischen Emiraten*, 4th edition, 2014, p. 41.

finance, investment, civil aviation, crude oil/natural gas as well as economic policy and finance are independently organized by each emirate on its own legislatorial and executorial responsibility.[34] Each emirate can pass its own laws, provided that such laws do not conflict with federal law of the UAE.[35]

The core principles of the legal system of the UAE are rooted in the Shari'ah. However, the legal system of the UAE mainly consists of a mix of Islamic law and European civil law concepts.[36] The legal concepts of the UAE have a common root in the Egyptian legal code, which was instituted in the late 19th / 20th centuries.[37] Since Egyptian law is strongly influenced by French law and includes many French civil law elements,[38] this explains the civil law influence in the UAE. While areas of social law, such as family law and inheritance law are primarily governed by the *Shari'ah*, areas of civil law, particularly business and commercial laws are in principle based on civil law system principles.[39] However, the legal system of the UAE also contains certain common law concepts and principles, since the territories of the UAE prior to the foundation of the UAE used to be under British control and the UAE have a strong exposure to common law through active commercial contact with common law jurisdictions.[40] In addition, many companies from the United States of America and Great Britain operate in the UAE, which leads to familiarity with common law principles in commerce.[41]

34 Seifert, *Rechtliche Rahmenbedingungen für Geschäftstätigkeiten in den Vereinigten Arabischen Emiraten*, 4th edition, 2014, p. 41.
35 Seifert, *Rechtliche Rahmenbedingungen für Geschäftstätigkeiten in den Vereinigten Arabischen Emiraten*, 4th edition, 2014, p. 41.
36 Khedr, *Overview of United Arab Emirates Legal System*, 2018, sec. 2.1, http://www.nyulawglobal.org/globalex/United_Arab_Emirates1.html,
37 Khedr, *Overview of United Arab Emirates Legal System*, 2018, sec. 2.1, http://www.nyulawglobal.org/globalex/United_Arab_Emirates1.html,
38 Khedr, *Overview of United Arab Emirates Legal System*, 2018, sec. 2.1, http://www.nyulawglobal.org/globalex/United_Arab_Emirates1.html,
39 Khedr, *Overview of United Arab Emirates Legal System*, 2018, sec. 3.5, http://www.nyulawglobal.org/globalex/United_Arab_Emirates1.html,
40 United Arab Emirates Business Guide 2014/2015, p. 82.
41 United Arab Emirates Business Guide 2014/2015, p. 82.

2. *Legal bodies*

a) The Federal Supreme Council

The Federal Supreme Council is the highest of the constitutional authorities of the UAE and is composed of the current rulers of the seven emirates.[42] Each emirate has a single vote.[43] Decisions of the Federal Supreme Council on substantive matters require a qualified majority of five council members including the votes of Abu Dhabi and Dubai.[44] Decisions on procedural matters can be taken by a simple majority vote.[45]

The Federal Supreme Council is, *inter alia*, responsible for the following matters:[46] formulation of general policy in all matters within the UAE;[47] ratification of various UAE laws before their promulgation, such as the Laws of the Annual General Budget and the Final Account;[48] ratification of decrees, if the constitution requires a ratification or the approval of the Federal Supreme Council[49]; and the election of the President and the Deputy to the President of the UAE.[50] Furthermore, the Federal Supreme Council has great influence on the appointment of judges of the Federal Supreme Court and the Federal Ministers.[51]

b) The Council of Ministers

The Council of Ministers of the UAE (also referred to as the Cabinet[52]) consists of citizens of the UAE known and chosen for

42 Art. 46 para. 1 sentence 2 Constitution of the United Arab Emirates.
43 Art. 46 para. 2 Constitution of the United Arab Emirates.
44 Art. 49 para. 1 sentence 1 Constitution of the United Arab Emirates.
45 Art. 49 para. 2 Constitution of the United Arab Emirates.
46 Art. 47 Constitution of the United Arab Emirates.
47 Art. 47 no. 1 Constitution of the United Arab Emirates.
48 Art. 47 no. 2 Constitution of the United Arab Emirates.
49 Art. 47 no. 3 Constitution of the United Arab Emirates.
50 Art. 51 Constitution of the United Arab Emirates.
51 United Arab Emirates, Business Guide 2014/2015, p. 83.
52 https://www.government.ae/en/about-the-uae/the-uae-government/the-uae-cabinet,

their competence and experience.[53] The Council of Ministers is composed of the Prime Minister, two Deputy Prime Ministers, the Ministers of the UAE and a General Secretariat, which assists the Council of Ministers in carrying out its work.[54] The Cabinet holds an ordinary meeting once a week, but the Prime Minister may convene an extraordinary session whenever he deems necessary, or at the request of a Minister.[55] The deliberations of the Council of Ministers are secret.[56] Its decisions are taken by majority vote of the members.[57] In the event of a tie, the Prime Minister's vote is decisive.[58]

The Council of Ministers is the constitutional manifestation of the executive.[59] It carries out all internal and external affairs of the UAE in accordance with the provisions of the constitution as well as the federal laws and is supervised by the President and the Federal Supreme Council.[60] One of the main tasks of the Council of Ministers is to initiate legislative drafts; it also supervises and discusses governmental acts.[61]

c) The Union National Council

The Union National Council is composed of 40 members. Abu Dhabi and Dubai each have eight seats, Sharjah and Ras al Khaimah each have six seats, Ajman, Umm al Qaiwain and Fujairah each have four seats.[62] In the past, all of the members of the Union National Council were directly appointed by the emirates, usually

53 Art. 55 Constitution of the United Arab Emirates.
54 https://government.ae/en/about-the-uae/the-uae-government/the-uae-cabinet,
55 https://government.ae/en/about-the-uae/the-uae-government/the-uae-cabinet,
56 Art. 61 para. 1 sentence 1 Constitution of the United Arab Emirates.
57 Art. 61 para. 1 sentence 2 Constitution of the United Arab Emirates.
58 Art. 61 para. 1 sentence 3 Constitution of the United Arab Emirates.
59 Art. 60 para. 1 Constitution of the United Arab Emirates.
60 https://government.ae/en/about-the-uae/the-uae-government/the-uae-cabinet,
61 Art. 60 para. 2 no. 2 Constitution of the United Arab Emirates; Art. 110 no. 2 lit. a Constitution of the United Arab Emirates; United Arab Emirates, Business Guide 2014/2015, pp. 83 et seq.
62 Art. 68 Constitution of the United Arab Emirates.

by a personal appointment by the ruler of the respective emirate.[63] However, in recent years a process has been initiated in order to increase the political participation of the citizen.[64]

The Union National Council has a Bureau consisting of a Chairman, a First and Second Vice-Chairman, and two supervisors.[65] The Council selects them from among its members.[66] The current Chairman and Speaker of the Federal National Council is Dr. Amal Al Qubaisi.[67] She is the first woman to hold this position.[68]

Deliberations of the Union National Council are invalid unless at least a majority of its members are present.[69] Decisions generally require an absolute majority of the votes of members present, except in cases where a special majority has been prescribed.[70] In the event of a tie, the chairman's vote is decisive.[71] In general, the sessions of the Union National Council are open for the public,[72] but they may be held behind closed doors at the request of a representative of the government, the Chairperson of the Council or three members of the council.[73]

The Union National Council is the main legislature organ of the UAE.[74] Federal draft laws prepared by the Council of Ministers first have to pass through the Union National Council for review and recommendations.[75] The Union National Council can amend

63 United Arab Emirates, Business Guide 2014/2015, p. 84.
64 United Arab Emirates, Business Guide 2014/2015, p. 84.
65 Art. 84 sentence 1 Constitution of the United Arab Emirates.
66 Art. 84 sentence 2 Constitution of the United Arab Emirates.
67 https://www.government.ae/en/about-the-uae/the-uae-government/the-federal-national-council-,
68 https://www.government.ae/en/about-the-uae/the-uae-government/the-federal-national-council-,
69 Art. 87 sentence 1 Constitution of the United Arab Emirates.
70 Art. 87 sentence 2 Constitution of the United Arab Emirates.
71 Art. 87 sentence 3 Constitution of the United Arab Emirates.
72 Art. 86 sentence 1 Constitution of the United Arab Emirates.
73 Art. 86 sentence 2 Constitution of the United Arab Emirates.
74 United Arab Emirates, Business Guide 2014/2015, p. 84.
75 Cf. Art. 110 no. 2 Constitution of the United Arab Emirates.

the original draft laws to adapt them to the requirements of citizens, or specific internal committees can draft and amend laws.[76]

d) The President of the UAE

The Supreme Council of the UAE elects from among its members a President of the UAE and a Deputy to the President of the UAE.[77] The term of office is five years and a reelection is possible.[78] The ruler of Abu Dhabi is traditionally the president of the UAE.[79] At the present time, this is H.H. Sheikh Khalifa bin Zayed Al Nahyan.[80] Since his election in 2004, he has initiated some major restructurings in the federal government and in the government of Abu Dhabi.[81] The current Vice-President and Prime Minister is H. H. Sheikh Mohammed bin Rashid Al Maktoum, who is also the ruler of Dubai.[82]

The president has various responsibilities. In particular, he is the Chairman of the Supreme Council and directs its debates;[83] he appoints the Chairman of the Council of Ministers,[84] and he signs the UAE laws, decrees and decisions, which the Supreme Council has ratified and promulgates them.[85] In addition, the president represents the UAE vis-à-vis other states and generally in all international relations.[86]

---

76 https://www.government.ae/en/about-the-uae/the-uae-government/the-federal-national-council-,

77 Art. 51 sentence 1 Constitution of the United Arab Emirates.

78 Art. 52 para. 1 sentence 1 and 2 Constitution of the United Arab Emirates.

79 https://www.kas.de/c/document_library/get_file?uuid=663ee5bf-b68d-0c6e-53b1-8b59174b35c4&groupId=286298,

80 https://www.government.ae/en/about-the-uae/the-uae-government/the-president-and-his-deputy,

81 https://www.uae-embassy.org/about-uae/about-government/his-highness-sheikh-khalifa-bin-zayed-al-nahyan,

82 https://www.government.ae/en/about-the-uae/the-uae-government/the-federal-national-council-,

83 Art. 54 no. 1 Constitution of the United Arab Emirates.

84 Art. 54 no. 5 Constitution of the United Arab Emirates.

85 Art. 54 no. 4 Constitution of the United Arab Emirates.

86 Art. 54 no. 9 Constitution of the United Arab Emirates.

3. Developments

The UAE are instantly developing the business laws to achieve their economic goals as set forth in the UAE Vision 2021. This particularly refers to the areas of foreign investment law, competition law and commercial arbitration.[87] The legislative process tries to include economic and social questions as well.[88]

In 2015, the UAE introduced a new Commercial Companies Law.[89] The Law contained numerous changes and introduced new concepts for the Limited Liability Company (LLC) and the Public Joint Stock Company (PJSC). These new concepts aim to meet the requirements of the international market and standards.[90] The modalities for setting up PJSCs and convening shareholders' meetings were partially liberalized.[91] For example, the minimum number of founders of a PJSC has been reduced from ten to five, which is particularly welcomed by family businesses and generally by LLCs considering an initial public offering.[92] In contrast, the rules governing the composition of the Board of Directors of PJSCs were further restricted: The majority of the Board members and the Chairman of the Board must be nationals of the United Arab Emirates. In addition, at least two thirds of the Board members must hold shares in the company.[93] In 2017, the Commercial Companies Law was amended by Federal Decree-

87 Cf. Khedr, *Overview of United Arab Emirates Legal System*, 2018, sec. 4.2.2.1 et seq, http://www.nyulawglobal.org/globalex/United_Arab_Emirates1.html,
88 Cf. Khedr, *Overview of United Arab Emirates Legal System*, 2018, sec. 4.2.2.1 et seq, http://www.nyulawglobal.org/globalex/United_Arab_Emirates1.html,
89 UAE Federal Law No. 2 of 2015; cf. Ulmer/Daghles, *Neue Regeln für Investoren in den Emiraten*, 2015, https://www.boersen-zeitung.de/index.php?li=1&artid=2015204067&artsubm=&subm,
90 Ulmer/Daghles, *Neue Regeln für Investoren in den Emiraten*, 2015, https://www.boersen-zeitung.de/index.php?li=1&artid=2015204067&artsubm=&subm,
91 Ulmer/Daghles, *Neue Regeln für Investoren in den Emiraten*, 2015, https://www.boersen-zeitung.de/index.php?li=1&artid=2015204067&artsubm=&subm=,
92 Ulmer/Daghles, *Neue Regeln für Investoren in den Emiraten*, 2015, https://www.boersen-zeitung.de/index.php?li=1&artid=2015204067&artsubm=&subm=,
93 Ulmer/Daghles, *Neue Regeln für Investoren in den Emiraten*, 2015, https://www.boersen-zeitung.de/index.php?li=1&artid=2015204067&artsubm=&subm=,

Law No. 18, particularly authorizing the Council of Ministers to grant foreign investors the right to aquire higher shareholdings in specific sectors.[94]

Furthermore, a new Foreign Direct Investment Law was issued in 2018.[95] The law in principle shall enable foreign investors to acquire shareholdings of more than 49% in companies domiciled in the UAE in certain sectors.[96] A higher investment was previously limited to free-zone registered companies.[97] With this new law a new FDI Unit and an FDI Committee formed by the Council of Ministers shall be formed, which shall suggest a positive list of certain economic sectors in which higher investment by foreign investors shall be admissible.[98] The investor interested in a 100% shareholding needs to apply for approval, and the Council of Ministers may impose restrictions on investors, such as only permitting an investment of less than 100%, or restricting the permit to one or a few emirates.[99]

In addition, recently Abu Dhabi has announced to reform its real estate law.[100] Foreign nationals shall be able to purchase and own land in Abu Dhabi investment zones.[101] Currently land ownership is restricted to citizens of the UAE or nationals of the GCC.[102] Foreign investors were restricted to a 99-year lease, a 50-year

---

94 http://www.wfw.com/wp-content/uploads/2018/11/WFWBriefing-UAE-Foreign-Direct-Investment-Law.pdf,

95 Federal Decree-Law No. 19 of 2018; https://www.biznet-consulting.com/federal-decree-law-no-19-of-2018-new-foreign-direct-investment-law-in-the-uae/,

96 http://www.wfw.com/wp-content/uploads/2018/11/WFWBriefing-UAE-Foreign-Direct-Investment-Law.pdf,

97 http://www.wfw.com/wp-content/uploads/2018/11/WFWBriefing-UAE-Foreign-Direct-Investment-Law.pdf,

98 https://www.dentons.com/en/insights/alerts/2018/november/13/new-foreign-direct-investment-law-for-

99 https://united-arab-emirates.taylorwessing.com/en/news/uae-foreign-direct-investment-law,

100 https://www.bayut.com/mybayut/new-property-law-abu-dhabi-2019/,

101 https://www.cnbc.com/2019/04/22/abu-dhabi-real-estate-law-reforms-are-gaming-changing-developer.html,

102 https://www.bayut.com/mybayut/new-property-law-abu-dhabi-2019/,

renewable lease, usufruct for a maximum of 99 years or a long-term leases for 25 years and more.[103]

4. *Gesellschafter und Partner der Dii*

The movement of Dii Desert Energy since 2009 could not have become a great long-term success without the support of numerous companies, institutions and governments. As Dii GmbH has been founded as an 'industry initiative' the companies that have played a direct role in Dii deserve to be mentioned explicitly. The list below shows chronologically the companies that have been connected with Dii as a shareholder or as a so called 'Associated Partner'.

103 https://www.bayut.com/mybayut/new-property-law-abu-dhabi-2019/,

Gesellschafter und Partner der Dii 2009

	Gesellschafter	*Partner*
1	ABB	Audi
2	Abengoa Solar	Bilfinger Berger
3	Cevital	Commerzbank
4	Desertec Foundation	Conergy
5	Deutsche Bank	3M EMEA
6	E.ON	Ersol Bosch Group
7	HSH Nordbank	Evonik
8	M+W Zander	First Solar
9	Solar Millennium	Flabeg Holding
10	Munich RE	Terna Energy
11	RWE	HSBC
12	Schott Solar	IBM
13	Siemens	Italgen
14		Kaefer
15		Morgan Stanley
16		Nur Energie
17		OMV
18		Saint Gobain
19		Schoeller Grpou
20		Tüv Süd

Gesellschafter und Partner der Dii 2010

	Gesellschafter	*Partner*	*Partner*
1	ABB	3M	
2	Abengoa Solar	AGC	Italgen
3	ACWA Power	Audi	Kaefer
4	Cevital	BASF	Lahmeyer International
5	Desertec Foundation	BearingPoint	Maurisolaire
6	Deutsche Bank	Bilfinger Berger	Max Planck Gesellschaft
7	Enel Green Power	Bosch	Morgan Stanley Bank
8	Eon	Commerzbank	Nur Energie
9	First Solar	Concentrix Solar	OMV Power International
10	Flagsol	Conergy	Scheaffler Technologies
11	HSH Nordbank	Deloitte	Schoeller
12	Munich RE	Dow Corning	SMA Solar Technology AG
13	M+W Group	Evonik	TÜV Süd
14	Nareva Holding	FCC Energia	Intesa Sanpaolo
15	Red Electrica	First Solar	
16	RWE	Flabeg	
17	Saint Gobain Solar	Fraunhofer Gesellschaft	
18	Schott Solar	GL Garrad Hassan	
19	Siemens	HSBC Bank PLC	
20	Terna	IBM	
21	Unicredit Group	ILF	

Gesellschafter und Partner der Dii 2011

	Gesellschafter	*Partner*	*Partner*
1	ABB	3M	Lahmeyer International
2	Abengoa Solar	AGC Group	Maurisolaire
3	ACWA Power	Audi	Max Planck Gesellschaft
4	Cevital	BASF	Morgan Stanley
5	Desertec Foundation	BearingPoint	Nur Energie
6	Deutsche Bank	Bilfinger Berger	OMV
7	Enel Green Power	Commerzbank	Bosch
8	E.ON	Concentrix Solar	Schaefller Technologies
9	Flagsol	Conergy	Schoeller Holding GmbH
10	HSH Nordbank	Deloitte	SMA Solar
11	Munich Re	DOW Corning	TÜV Süd
12	M+W Group	Evonik	Intesa Sanpaolo
13	Nareva Holding	FCC Energia	
14	Red Electrica	First Solar	
15	RWE	Flabeg	
16	Saint Gobain Solar	Fraunhofer Gesellschaft	
17	Schott Solar	GL Garrad Hassan	
18	Siemens	HSBC	
19	Terna	IBM	
20	Terna Energy	ILF	
21	Unicredit Gorup	Italgen	
22		Kaefer	

Gesellschafter und Partner der Dii 2012

	Gesellschafter	*Partner*	*Partner*
1	ABB	AGC Group	Maurisolaire
2	Abengoa	Audi	Max Planck Gesellschaft
3	ACWA Power	BASF	Morgan Stanley
4	Cevital	BearingPoint	Nur Energie
5	Desertec Foundation	Bilfinger	OMV
6	Deutshbank	Commerzbank	Bosch
7	Enel Green Power	Conergy	Schoeller
8	Eon	Dow Corning	Shell
9	First Solar	Evonik	SMA Solar
10	HSH Nordbank	FCC Energia	Soitec
11	Flagsol	First Solar	TUV Süd
12	Munich RE	Flabeg	3M
13	M+W Group	Fraunhofer Gesellschaft	
14	Nareva Holding	GL Garrad Hassan	
15	Red Electrica	HSBC	
16	RWE	IBM	
17	Saint Gobain Solar	ILF	
18	Schott Solar	Intesa Sanpaolo	
19	Siemens	Italgen	
20	Terna	Kaefer	
21	Terna Energy SA	Lahmeyer	
22	Unicredit Group	Leoni	

Gesellschafter und Partner der Dii 2013

	Gesellschafter	*Partner*
1	ABB	AGC
2	Abengoa	Audi
3	ACWA Power	BeairngPoint
4	Deutsche Bank	Bilfinger Berger
5	Enel Green Power	ILF
6	Eon	Intesa Sanpaolo
7	First Solar	Italgen
8	HSH Nordbank	Lahmeyer International
9	MUNICH RE	Leoni
10	Nareva Holding	Maurisolaire
11	RED Electrica	Max Planck - Gesellschaft
12	RWE	Shell
13	Schott Solar	SMA
14	Saint Gobain Solar	Soitec
15	Terna Energy SA	TUV SUD
16	Unicredit	
17	SGCC	

Gesellschafter und Partner der Dii 2014

	Gesellschafter	*Partner*
1	Abengoa	Audi
2	ACWA Power	BearingPoint
3	Deutsche Bank	Bilfinger Berger
4	Enel Green Power	ILF
5	E.ON	Mondeagon
6	First Solar	Intesa Sanpaolo
7	Munich Re	Italgen.
8	RED Electrica	Lahmeyer International
9	RWE	Leoni
10	Schott Solar	Maurisolaire
11	Terna Energy SA	Shell
12	Unicredit	TÜV Süd
13	SGCC	

Gesellschafter und Partner der Dii 2015

	Gesellschafter	*Partner*
1	RWE	5 Capitals
2	SGCC	Aalborg
3	ACWA Power	ABB
4		Acciona
5		Building Energy
6		CNIM
7		Engie
8		Esolar
9		First Solar
10		Frenell
11		Kaefer
12		Sargent & Lundy
13		Siemens
14		Solar Reserve
15		Maurisolaire
16		Terranex
17		TSK
18		Worley Parsons
19		Yingli Solar

Gesellschafter und Partner der Dii 2016

	Gesellschafter	*Partner*
1	innogy	5 Capitals
2	SGCC	ABB
3	ACWA Power	Al Fanar
4		Alchimede
5		ATA
6		Aew Truepower
7		Cobra
8		Diseprosa
9		Effergy Energia
10		Energynest
11		First Solar
12		GE
13		GWI/MAL
14		ILF
15		Max Planck Gesellschaft
16		Mondragon Education
17		Roland Berger
18		SANK ENERJI
19		SBP
20		Siemens
21		Solar Reserve
22		Maurisolaire
23		TSK
24		Worley Parsons

Gesellschafter und Partner der Dii 2017

	Gesellschafter	*Partner*
1	ACWA Power	Al Fanar
2	innogy	Al Gihaz
3	SGCC	ATA
4		Belectric
5		Cobra
6		Diseprosa
7		First Solar
8		ILF
9		Mondragon
10		Noor Solar Technology
11		SBP
12		Siemens
13		SUNTECH
14		Maurisolaire
15		Thyssen Krupp
16		TSK
17		Worley Parsons
18		Amana
19		Europagrid
20		ABB
21		Nivigant
22		Roland Berger

Gesellschafter und Partner der Dii 2018

	Gesellschafter	*Partner*
1	ACWA Power	ABB
2	Innogy	Access
3	SGCC	Al Gihaz
4		Amana
5		Belectric
6		Cobra
7		Enara
8		Enerray
9		Europa Grid
10		ILF
11		Krinner
12		Mondragon
13		Nivigant
14		Noor Solar Technology
15		Siemens
16		Sterling and Wilson
17		Suntech
18		Maurisolaire
19		Thyssen Krupp
20		TSK
21		Worley Parsons
22		First Solar

Gesellschafter und Partner der Dii 2019

	Gesellschafter	*Partner*	*Partner*
1	ACWA Power	Siemens	Worley Parsons
2	innogy	Enerray	SQM
3	SGCC	Enerwhere	Sterling & Wilson
4		Envirofina	Roland Berger
5		Cobra	Mauri Solair
6		TSK	ILF
7		al Gihaz	
8		EuropaGrid	
9		Nivigant	
10		Ecolog	
11		Neom	
12		E-Nara	
13		ThyssenKrupp	
14		ABB	
15		Amana	
16		Yellow Door	
17		Meyer Burger	
18		Suntech	
19		Krinner	
20		Masdar	
21		Mondragon	
22		Nomad	

Acknowledgements

We would like to thank all the people who have made it possible for us to write this book and who have often personally inspired us to do so. First and foremost, the many people who have worked very hard for Dii and our mission: no emissions. Next, the numerous governments and parliamentarians of the participating countries, the supporting companies, institutions and many partners and friends of Dii Desert Energy. Without them, the extensive knowledge processes on the topics of desert electricity and emission-free energy would not have been possible in and from the desert countries of the MENA region. A wide circle of prominent people confirmed to us that Desertec is more alive than ever before and that the deserts of our earth will undoubtedly become an integral part of the global energy supply! In particular, our thanks go to our families who went through thick and thin with us and were always ready for critical discussion.

Paul van Son and Thomas Isenburg

Literature

Dii Publications via website

Anger H., Neurer D. (2009): Deutschland bei Wüstenprojekten im Vorteil, in: Handelsblatt, 14.07.2009.

Baumer A., Kepler M. (2017): Der Niedergang der deutschen Solarwirtschaft, in: Augsburger Allgemeine, 17.5.2017.

BBC (2015) Should we solar panel the Sahara desert? London

Bebel A. (1879): Frau im Sozialismus, Zürich.

Bebel A. (1910): Woman and Socialism, 50. anniversary edition, New York.

Balser M., Fromm Z. (2009): Wüstenstrom für Deutschland, in: Süddeutsche Zeitung, 16.06.2009.

Balser M. (2013): Desertec Stiftung steigt aus Wüstenstrom-Projekt aus, in: Süddeutsche Zeitung, 30.06.2013.

Balser M. (2013): Schatten über Sawian, in: Süddeutsche Zeitung, 27.06.2013.

Brown T.W. et. Al.: Response to'Burden of proof: A comprehensive review of the feasibility of100% renewable-electricity systems' in: Renewable and Sustainable Energy Reviews 92 (2018) 834 - 847

Calderbank S. (2013): Desertec abandons Sahara solar power export dream, in Euractiv, 31.5.2013

Celizalilla A., Wiebelt M., Blohmke J., Klepper G. (2014): Desert Power 2050: Regional and sectoral impact of renewable electricity production in Europe, the Middle East and North Africa, in: Kiel Working Paper No. 1891, January 2014.

Desertec Industrial Initiative (2009): Vision, Mission and Objectives of the Desertec Industries Initiative (Dii), Berlin.

Desertec-UK (2013): http://www.desertec-uk.org.uk

Dettmann K.-D., Heuk K., Schulf D. (2013): Elektrische Energieversorgung, Wiesbaden.

Deutsche Gesellschaft Club of Rome (2009): Clean power from Deserts – The DESERTEC Concept for Energy Water and Chemicals and Climate Security, Bonn.

Deutsche Gesellschaft Club of Rome (Hrsg.) (2011): Der Desertec-Atlas: Weltatlass zu den erneuerbaren Energien, Hamburg.

Deutsche Industrie- und Handelskammer in Marokko – DIHK (Hrsg.) (2016): Zielmarktanalyse Marokko: "Windenergie (Technologien zur Eigenversorgung/Zulieferindustrie", Casablanca.

Deutsches Zentrum für Luft- und Raumfahrt (DLR) (2005): MED-CSP Studienbericht: Solarthermische Kraftwerke für den Mittelmeeraum, Stuttgart.

Deutsches Zentrum für Luft- und Raumfahrt (DLR) (2006): TRANS-CSP Studienbericht: Trans-Mediterraner Solarstromverbund, Stuttgart.

Edison A. E., (1931): https://www.filos-invest.de/direktinvestments/direktinvestment-photovoltaik/

Dii GmbH (Hrsg.) (2012a): Desert Power 2050 – Perspectives on a Sustainable Power System for EUMENA, München

Dii GmbH (Hrsg.) (2012b) Vision, Mission and Objectives of Dii GmbH

Dii GmbH (Hrsg.) (2013): Desert Power: Getting started, München

Dii GmbH (Hrsg.) (2014): Desert Power: Getting connected, München

Dii GmbH (Hrsg.) (2017): Who's Who, Netzwerkführer der Dii, Dubai.

Dittmann F. (2012): Frank Shumann und die frühe Nutzung der Solarenergie, in: Ingenieure in der technokratischen Hochmoderne, Münster.

dpa (2009): Töpfer und Grüne loben Wüstenstrom-Projekt, in: Focus online, 13.07.2009.

dpa (2012): Deutsche Firmen ziehen sich aus Desertec zurück, in: Spiegel online, 13.12.2009.

Dubessy F., (2010) Le projet Transgreen devient la société

Medgrid, in Econ sturm, 10.10.2010

Ehlerding, S. (2018): Donald Trumps holpriger Abschied vom Pariser Abkommen, in : Tagesspiegel, 01.06.2018.

FIZ Karlsruhe GmbH – Leibniz-Institut für Informationsinfrastruktur (Hrsg.) (2013): Solarthermische Kraftwerke. Konzentriertes SOnnenlicht zur Energieerzeigung nutzen. Bonn.

Franken F. (2013): Bericht aus der Zukunft – Wie der der grüne Wandel funktioniert, München.

Gasch R., Telwe J. (2013): Windkraftanlagen, Heidelberg.

Günther H. (1931) In hundert Jahren, Stuttgart

Heckel M., (2011): Desertec – oder der Traum von der unendlichen Energie, Potsdam.

IPAMED (2010): http://www.ipemed.coop/en/events-r18/mediterranean-breakfast-meetings-c53/solar-projects-in-the-mediterranean-progress-and-questions--a313.html

Gassmann, M. (2009): In die Wüste geschickt, in: Financial Times Deutschland, 30.10. 2009.

Großmann J. (2012): Photovoltaik in Deutschland macht ökonomisch so viel Sinn wie Ananaszüchten in Alaska, Tagungsbeitrag auf der Handelsblatt-Tagung am 18. Januar 2012.

Hourani A. (2016): Die Geschichte der arabischen Völker, Frankfurt.

Hoffmann K. P. (2013): Desertec feuert die Geschäftsführerin, in: Tagesspiegel, 9.7.2013.

Hoffmann K. P. (2013): Ich fühle mich wie der Präsident einer Bananen Republik, in: Tagespiegel, 8.12.2013.

Herrmann R. (2011); Strom für Europa aus der Wüste Marokkos, in: Frankfurter Allgemeine Zeitung, 03.11.2011.

Hoffmann A. R., Varadi P. F., Wouters F. (2018) The Sun Is Rising in Africa and the Middle East: On the Road to a Solar Energy Future (Pan Stanford Series on Renewable Energy, Band 9) Singapore

IRENA (2018): Climate-Safe Energy: Competes on Cost Alone, Abu Dhabi

IRENA (2018): Renewable Energy and Jobs – Anual Review 2018, Abu Dhabi

Imwinkelried D. (2011): Fünfzehn Prozent des Stroms aus der Wüste, in: Neue Züricher Zeitung, 18.04.2011.

Isenburg T., Itasse S. (2014): Wüstenprojekte sind auf dem Weg zu neuen Rekorden, in: MM Maschinenmarkt, 27.01.2014.

Isenburg T. (2016): Internationale Projekte richtig managen, in: KeNext, 03.03.2016.

Isenburg T., (2016): Den Mangel überwinden, in: Energiewirtschaftliche Tagesfragen 2016, Heft 1/2.

Isenburg T., (2019) Große Ideen für ein mit Energiereichtum gesegnetesLand, in Wasserkraft & Energie 2/2019

Lahdiri C. (2011): Energies: Vers un accord de coopération entre Desertec II et l'Algérie en décembre , in: Algeria-Watch , 15.11.2011.

Lubbadeh J., (2009): Superkraftwerk Sonne, in: Spiegel Online 01.12.2009.

Lüders M. (2015): Wer den Wind sät, München.

Manager Magazin (2009): Einstieg in Solarprojekte am Mittelmeer, in: Manager Magazin, 10.07.2009.

Marshall T. (2017): Die Macht der Geographie, Köln.

Martin M. (2016): Planet Wüste, München.

Meadows D., Meadows, Donella H., Randers, Jorgen (1973): Die Grenzen des Wachstums. Bericht des Clube of Rome zur Lage der Menschheit, Hamburg.

Mouchat A. (1877): Die Sonnenwärme und ihre industrielle Anwendung, Reprint und deutsche Übersetzung 1987, Amsterdam.

Neidlein H.-C. (2008): Kraftwerke und Stromnetze, in: vdi-Nachrichten ,18.04.2008.

N.N. (1916): Uses Egypts Sun for Power, in: The New York Times, 02.11.1916.

Olah G. A., Goeppert A., Prakash G.K.S. (2018): Beyond Oil and Gas: The Methanol Economy, Weinheim.

Petermann J. (2008): Sichere Energie im 21. Jahrhundert, Hamburg.

Pflüger F. (2009): Sahara Strom bringt Afrika und Europa zusammen, in: Spiegel Online, 1.7.2009.

Rahmstorf S., Schellnhuber H.J. (2007): Der Klimawandel. Diagnose, Prognose, Therapie, München

Richtlinie 2009/28/EG (Erneuerbare-Energien-Richtlinie) (2009)

Rößler W. (2009): Eine kleine Nachtphysik, Große Ideen und ihre Entdecker, Reinbeck bei Hamburg.

Roussel F. (2010): Transgreen, Desertec : deux projets pour un

même but, actu-enviroment, 11.6.2010

Sasse D. (2006): Franzosen, Briten und Deutsche im Riffkrieg 1921-1926, Berlin.

Scheer, H. (1999): Solare Weltwirtschaft: Strategie für die ökologische Moderne, München.

Scheer H. (2010): Der energethische Imperativ, München.

Scholl-Latour P. (2011): Arabiens Stunde der Wahrheit, Berlin.

Schroeder J. (2004): Im Rausch der Erkenntnis, in: Geo Epoche 12/2004.

Stolten D., Scherer V. (Hrsg.) (2013): Transition to Renewable Energy Systems, Weinheim.

Uken M., (2012): Der Wüstenstrom bekommt neue Freunde, in: Zeit online, 08.11.2012.

Uken M., (2013) Der Subventionswahn ist ungebrochen, in: Zeit online 14.10.2013.

Van Son P., Ruderer D. (2015): Capturing Synergies Among the Power Markets Around the Mediterranean, Robert Schuman Centre for Advanced Studies Research Paper No. RSCAS 2015/42.

Von Hiller C. (2012): Erstes Desertec-Projekt entsteht in Tunesien, in: Frankfurter Allgemeine Zeitung, 24.01.2012.

Wagemann H.-G., Eschrich H. (2007): Photovoltaik, Wiesbaden.

Wachstum, Bildung, Zusammenhalt. Der Koalitionsvertrag zwischen CDU, CSU und FDP für die 17. Legislaturperiode, Berlin 2009.

Wijk A. v. et. Al. (2018) Hydrogen, the key to the energy transition, Delft

Wijk A. v. et. Al. (2017): Solarpower to the people, Amsterdam

Wandler R. (2002): Deutsch vergast, marokkanisch vergessen, in: TAZ 26.01.2002.

Weishaupt G. (2010): Die Hoffnungsträgerin vieler Dax Chefs, in: Handelsblatt, 26.10.2010.

Wissenschaftlicher Beirat der Bundesregierung Globale Umweltveränderungen (2003): Welt im Wandel: Energiewende zur Nachhaltigkeit, Berlin.

Wetzel D. (2013): Saudi Arabien unterstützt Umsetzung von Desertec, in: Welt, 19.02.2013.

Wolf C., (2013): Stolpersteine auf dem Weg zur Realisierung von Desertec, HWWI Insights, 05/2013

List of abbreviations

AHK	Auslandshandelskammern (Chambers of Commerce Abroad)
ANME	Agence Nationale pour la Maîtrise (National Energy Agency Tunesia)
BMZ	Bundesministerium für Wirtschaftliche Zusammenarbeit und Entwicklung (Federal Ministry for Economic Cooperation and Development)
BMUB	Bundesministerium für Umwelt, Naturschutz und Reaktorsicherheit (Federal Ministry for the Environment, Nature Conservation and Nuclear Safety)
BND	Bundesnachrichtendienst
CCS	Carbon Capture and Storage
CEPRI	China Electric Power Research Institute (subsidiary of SGCC)
COP21	Paris Climate Agreement (2015)
COP22	World Climate Change Conference in Marrakech (2016)
CSP	Concentrated Solar Power / Solar Thermal power production
DESY	Deutsches Elektronen-Synchrotron
DEWA	Dubai Electricity and Water Authority

Dii	originally ‚Desertec Industrial Initiative' / brand-name: Dii Desert Energy
DLR	Deutsches Zentrum für Luft- und Raumfahrt (German Aerospace Center)
DP2050	Dii Report Desert Power 2050
EIB	European Investment Bank
EBRD	European Bank for Reconstruction and Development
ENTSO-E	European Grid Operators
EPC	Engineering, Procurement and Construction
EU-MENA	the countries of Southern Europe, North Africa and the Middle East
FLN	National Liberation Front (Algeria)
GCC	Gulf Cooperation Council
GCCIA	Gulf Cooperation Council Interconnection Authority
GEIDCO	Global Energy Interconnection Development & Cooperation Organization
Ghorfa	Arab-German Chamber of Commerce and Industry
GIS	Geo Information Systems

GIZ	Gesellschaft für Internationale Zusammenarbeit (Society for International Co-operation)
GoO	Guarantee of Origin
GW	gigawatt
HFO	Heavy Fuel Oil
IEA	International Energy Agency
IFC	International Finance Corporation
IPCC	International Panel of Climate Change
I-REC	International Renewable Energy Certificates
IREC	Interstate Renewable Energy Council
IRENA	International Renewable Energy Agency
IS	Islamic State
ISI	Institut für System- und Innovationsforschung (Institute for Systems and Innovation Research)
KfW	Kreditanstalt für Wiederaufbau (Reconstruction Loan Corporation)
KSA	Kingdom of Saudi Arabia
kWh	kilowatt-hour
LAS	League of Arab States

LNG	Liquefied Natural Gas
MASEN	Moroccan Agency for Solar Energy
MEDGRID	Mediterranean Grid Studies
MED-TSO	Mediterranean Transmission System Operators
MENA	the Middle East and North Africa
MESIA	Middle East Solar Industry Association
MIT	Massachusetts Institute of Technology
MW	megawatt
NEOM	planned smart mega city in Saudi Arabia
NERC	Jordan National Energy Research Centre
NGO	Non-governmental organisation
IPP	Independent Power Producer
OECD	Organisation for Economic Cooperation and Development
ONEE	Office National de l'Electricité et de l'Eau Potable (National Office for Electricity and Potable Water, Morocco)
PPP	Public-Private Partnership
PPA	Power Purchase Agreement

ppm	parts per million
PV	photovoltaic
RAROC	Risk-Adjusted Return on Capital
RCREEE	Regional Centre for Renewable Energy and Energy Efficiency (Cairo)
RECS /iRECs	Renewable Energy Certificate System
SEC	Southern California Edison
SGCC	State Grid Corporation of China
STEG	Tunisian energy supplier
TREC	Trans-Mediterranean Renewable Energy Cooperation
UAE	United Arab Emirates
UDMA	Union of the Algerian Manifesto
UfM	Union for the Mediterranean
UNEP	United Nations Environment Programme
VDI	Verein Deutscher Ingenieure (Society of German Engineers)